Class, Individualization and Late Modernity

Titles include:
Will Atkinson
CLASS, INDIVIDUALIZATION AND LATE MODERNITY
In Search of the Reflexive Worker

Ben Rogaly and Becky Taylor
MOVING HISTORIES OF CLASS AND COMMUNITY
Identity, Place and Belonging in Contemporary England

Margaret Wetherell (*editor*)
IDENTITY IN THE 21ST CENTURY
New Trends in Changing Times

Margaret Wetherell (*editor*)
THEORIZING IDENTITIES AND SOCIAL ACTION

Class, Individualization and Late Modernity

In Search of the Reflexive Worker

Will Atkinson
University of Bristol, UK

First published 2010 by
PALGRAVE MACMILLAN

Palgrave Macmillan in the UK is an imprint of Macmillan Publishers Limited, registered in England, company number 785998, of Houndmills, Basingstoke, Hampshire RG21 6XS.

Palgrave Macmillan in the US is a division of St Martin's Press LLC, 175 Fifth Avenue, New York, NY 10010.

Palgrave Macmillan is the global academic imprint of the above companies and has companies and representatives throughout the world.

Palgrave® and Macmillan® are registered trademarks in the United States, the United Kingdom, Europe and other countries.

ISBN 978–0–230–24200–5   hardback

This book is printed on paper suitable for recycling and made from fully managed and sustained forest sources. Logging, pulping and manufacturing processes are expected to conform to the environmental regulations of the country of origin.

A catalogue record for this book is available from the British Library.

Library of Congress Cataloging-in-Publication Data

Atkinson, Will, 1983–
    Class, individualization, and late modernity : in search of the reflexive worker / Will Atkinson.
        p. cm.—(Identity studies in the social sciences)
    Includes bibliographical references.
    ISBN 978–0–230–24200–5
    1. Social classes.   I. Title.
        HT609.A85 2011
        305.5—dc22

2010027490

Printed and bound in Great Britain by
CPI Antony Rowe, Chippenham and Eastbourne

For Tant and Vikki

Contents

Acknowledgements

My gratitude to all those who took part in the research, and to Harriet Bradley, Ashley Cooper, Stuart Davies, Esther Dermott, Jackie Friel, Lydia Hayes, Jo Haynes, Noreen Kelly, Ruth Levitas, Laura Merlino, Mike Savage, Maggie Studholme, Paula Surridge and the series editors for their various inputs. Gregor McLennan deserves extra thanks for being an unfailing source of support, encouragement and constructive criticism throughout.

1
Introduction: From Affluence to Reflexivity

Forty years ago, the spectre of embourgeoisement haunted the sociology of class. Ever-increasing affluence, relative parity of incomes and living conditions and the expanded availability of consumer goods had all, so proponents of this famous thesis asserted, ensured the cultural and political assimilation of the working class into the middle rungs of society and, as a consequence, effectively rendered the concept of class redundant (see, for example, Zweig, 1961). Lifestyles and social values had converged, the argument went, with the erstwhile working class eagerly appropriating the tastes and leisure pursuits of the growing middle class, unapologetically jettisoning their once unbreakable commitment to collectivism and trade unionism in favour of bourgeois privatism, individualism and status-obsession and turning to Conservatism in the political arena as the only force capable of ensuring the maintenance of their new-found principles. In sum, no distinguishable difference warranting sociological attention existed between occupational groups, it was claimed, and despite its scant empirical backing this idea soon accrued considerable popular purchase.

Enter John Goldthorpe, David Lockwood, Frank Bechhofer and Jennifer Platt, probably the most celebrated research team in the history of British sociology. After promptly taking to task representatives of the embourgeoisement thesis for their conceptual inadequacies and attempting to reformulate their propositions as sound hypotheses (Goldthorpe and Lockwood, 1964), Goldthorpe et al. produced three slim and sober volumes of empirical research systematically demolishing their claims (Goldthorpe et al., 1968a, 1968b, 1969). Having interviewed 229 manual workers in three separate industrial plants in Luton, the *Affluent Worker* team, as they came to be known, concluded that, while there had indeed been profound changes in the working class in the post-war

1

period, they could hardly be characterized as embourgeoisement. Yes, there was a degree of 'normative convergence' between sections of the working class and white-collar workers, but this revealed not the dissolution of that presumed-dead class but its transformation into a new, privatized collection of workers characterized by an instrumental approach to work, a 'family centredness' and a view of societal stratification that fixed on economic differences (see also Lockwood, 1966). They may have had a novel form, in other words, but distinct class divisions and experiences remained.

Four decades on, however, the social world has changed and, though the vanquished apparition of embourgeoisement has been safely buried by the sands of time, the concept of class is under threat once again. Pushing to the forefront of an obstreperous multitude of perspectives critical of the veteran sociological tool is a new challenge – one with consequences just as stark as embourgeoisement but anchored in the global socio-political climate of the late twentieth century, one that has theorized recent transformations in the social landscape without capitulating to the excesses of postmodern proclamations, and one that commands wide influence and discussion within the academy while also existing in simplified form in the political arena. This challenge is that posed by the various theories putting *reflexivity* at their core, that is to say the thesis of 'individualization', advanced in slightly different versions by German sociologist Ulrich Beck and Polish émigré Zygmunt Bauman, and the kindred ideas on 'late' or 'high' modernity forwarded by British social theorists Anthony Giddens and Margaret Archer. All four thinkers, in diverse ways and with differing degrees of equivocation, posit the withering of tradition and the onset of new social forces prising individuals from their old collective modes of existence, both of which ensure that people no longer have any choice but to *actively think and choose* how to live, what to value and what to become. Class has ceased to constrain or enable life decisions, they claim, and no longer produces taken-for-granted ways of living that shape behaviour, values, views and identities. In short, citizens of contemporary Western societies are said to be not merely affluent workers, but *reflexive workers* living in an environment of unmitigated choice.

The ability of class to explain patterns of difference and inequality is thus once more put into question, its position in the sociological armoury queried and its utility for understanding the key political issues of our time thrown into doubt. Yet surprisingly there has, to date, been no attempt among those faithful to the concept to provide a substantial assessment of these ideas. Sure enough they have rebuked Beck and the others, dismissed them or undertaken partial examinations of their claims while continuing to produce theoretical and empirical contributions to their topic, but no sustained head-on assessment has yet emerged. Furthermore, while we may well note the *theoretical* incoherence of each position (Atkinson, 2007a, 2007b, 2008), to the extent that we can safely dispense with the possibility of accepting any of them *en bloc*, conceptual critique is not enough on its own to fully invalidate the theories of reflexivity. Some of the observed empirical trends and themes could, after all, be occurring and even have profound ramifications for class, but have so far been clad in an insufficient conceptual vocabulary that confuses more than it clarifies. Once again, therefore, it is essential to adjudicate the fate of class in Britain by introducing the ultimate arbiter: empirical research. Thus, rekindling the spirit of the *Affluent Worker* research, with a due nod to Gallie's (1978) hunt for the 'new working class', the present study goes in search of the reflexive worker.

Before detailing how it intends to proceed, the socio-historical and intellectual scene needs to be set in a little more detail. If we are to understand how and why these arguments have emerged and gathered momentum, why they pose such a challenge to class analysis and the failure of class researchers to provide a definitive assessment, then we must delve into the social transformations that gripped late-twentieth century Western societies and the changes in and challenges to research on the most contested of all sociological concepts.

Fin-de-siècle class analysis

Let us begin in the 1980s, when sociological research on class was dominated by two perspectives which, despite diverging in one key respect, nevertheless shared a collection of methodological and conceptual tenets that set them apart from others. These were, on the one hand, the 'analytical Marxism' of American sociologist Erik Olin Wright (1978, 1985) and, on the other, the Nuffield programme of social mobility research associated above all with John Goldthorpe (1980, 1987). Though falling on either side of the long-running schism in class analysis between those marching under the banner of Marxism and those taking inspiration from the writings of Weber, both (though Goldthorpe more so) tended to prioritize *empirical research* over theory building, conducting the latter – often in the form of trying to establish the 'correct' boundaries between classes or to explain concrete findings – only to serve the former or when provoked by criticism, and both pursued exclusively *quantitative analysis* of large-scale datasets employing often complex statistical techniques. These two facets were significant in establishing their influence: they broke with the prevailing approaches to class – theoretical musing, or historical or qualitative research – and offered for the first time, in different forms, the possibility of conducting large-scale comparative quantitative analysis based on rigorously constructed and theoretically grounded class categories rather than, as before, gradational scales operating with a simplistic conception of 'classes' ultimately reducible to the division between manual and non-manual workers. They are, therefore, by no means detractions, but their one-sidedness and sometimes extreme manifestation attracted repeated accusations of theoretical attenuation or empiricism (in Goldthorpe's case at least – Pahl, 1993; Morris and Scott, 1996) and, so some claimed, ultimately threatened to isolate class analysis from broader themes in social theory and, indeed, from any sociologist unfamiliar with the arcane language of advanced statistical procedures (Crompton and Scott, 2000; Savage, 2000).

Furthermore, both Wright and Goldthorpe focused on the delineation of objective, static class *structures* as matrices of independent variables and the examination of their consequences (for income, mobility and such like) and, in following through on this, both, like their classical progenitors, restricted classes to the *economic* domain by conceiving them as aggregates of occupations differentiated by certain production- or market-based characteristics. The two have therefore not only wrestled repeatedly, and for many unsatisfactorily, with how best to account for those not actually in employment but, coupled with their substantive interests in the numbers in each class position or with the patterning of differential life chances, marginalized issues of history, culture, subjectivity and identity from the conceptualization of class beyond the study of voting patterns or quantifiable indicators of class identification or consciousness (criticized at length in Marshall, 1988; Wright et al., 1989; Emmison and Western, 1990; Fantasia, 1995). Finally, both, as a complement to their large-scale, quantitative and economistic orientation, turned to *utilitarian* models of human agency – in Wright's (1985) case, sitting uncomfortably with his lapses into determinism or his lip service to the importance of the 'lived experience' of class in other places (see contributions to Wright et al., 1989) and absent in more recent works, Roemer's (1982) Marxist reworking of game theory; for Goldthorpe (1991), who had left unelaborated his theoretical position on agency for some time, a variety of rational action theory he was to develop and defend at greater length in later years (2007a, 2007b).

Changes and challenges at a century's end

This is not to say that there were no alternatives: the Cambridge school, who see social stratification as emergent from interaction patterns, had a small but loyal band of followers, and cultural or qualitative forms of Marxism, such as that pursued by Michael Burawoy (1979), were constantly battling with Wright's perspective. Still, there was no doubt that, by the early 1990s, Wright's and Goldthorpe's programmes were flourishing and the field of class analysis rigidifying around their partial opposition, with the comparison and validation of the different schemes and debates over the adequacy of particular statistical procedures becoming standard fare.

Ironically, however, exactly the same period was marked by an abundance of intertwined social processes in Western societies prompting a tide of declarations that class was an irrelevant and anachronistic concept. Such claims are hardly new, admittedly – sociologists of class have had to fend off a succession of challengers since they first carved out their subject area, whether they were the theorists of embourgeoisement encountered above, heralds of post-industrialism or thinkers of the New Left disappointed by the pacified working class. Yet the assertions accompanying the social changes of the last thirty years or so that supply the intellectual backcloth and bedfellows of individualization and reflexivity have been more sustained, multifaceted, forceful and explicit than ever before, often constituting component parts of forthright announcements that a novel period of social history necessitating a clear-out of the conceptual cupboards has been set in motion. For the sake of expositional clarity they can be categorized according to the key sphere upon which they bear – the economy, culture or politics – and taken in turn.

Beginning with the economic realm, then, the continued shrinkage of the manufacturing and extractive sectors in the West has, some argue, begun to erase from the occupational landscape the archetypal industrial proletariat as imagined by Marx. In societies structured around the production of knowledge, with a steady growth in the provision of and numbers passing through post-compulsory education, or the provision of services from financial consultancy to low-paid personal services like hairdressing and couriering, in its wake stand a heterogeneous batch of workers with economic stability and skills of some kind on the one hand and an un(der)employed and dispossessed mass on the other (Gorz, 1982; Lash and Urry, 1987, 1994; Pahl, 1989; Pakulski and Waters, 1996: 57–8; Gray, 1998). Not only that, numerous commentators claim, but, alongside this post-industrial turn, 'advanced' Western economies today no longer operate principally with the production system popularized by Henry Ford or the scientific management of Frederick Taylor but instead, in adapting to heightened global competition, technological developments and the *laissez-faire* disposition of neo-liberal governments, exhibit new features that can only be described as 'post-Fordist' (see Hall and Jacques, 1989; Harvey, 1989; Kumar, 1995).

The key notion here is *flexibility*: in terms of production, mass production is exchanged for pliable niche production; in terms of organizational structure, hierarchies are flattened and managerial functions streamlined; and in terms of the workforce, 'jobs for life' are dissolved as contracts from top to bottom of the occupational ladder become temporary and insecure in a drive to improve efficiency. All this, the argument goes, demolishes working-class collectivism and the strength of the trade unions (Harvey, 1989) while working to ensure that unemployment and poverty – and hence life chances *à la* Max Weber and Goldthorpe – are no longer distributed according to the class mould (see Leisering and Leibfried, 1999). In fact, critics of class claim, coupled with the vast differentiation of occupations, extensive state services and increased affluence and spare time of today the sphere of work or production as a whole – the domain of classes as traditionally conceived – is simply of less relevance in maintaining domination or determining attitudes, identities, lifestyles and life chances than other social divisions or the realm of leisure and consumption (Alt, 1976; Bell, 1976; Offe, 1985; Saunders, 1987; Pahl, 1989; Pakulski and Waters, 1996; Baudrillard, 2001).

But is the sphere of leisure and consumption, or, more broadly, culture, not simply home to class differences translated into consumer goods and lifestyles? Pointing to all manner of contemporary features of Western societies, the critics' answer is a resolute 'no' (see *inter alia* Lash and Urry, 1987, 1994; Harvey, 1989; Clark and Lipset, 1991; Featherstone, 1991; Crook et al., 1992; Baudrillard, 2001). The improved income of manual workers, for example, has opened up access to all manner of goods and lifestyle pursuits previously unattainable – all but the most disadvantaged can afford wide-screen televisions, exotic holidays and all the other symbols of affluence now. With the conspicuous arrival of postmodernism, furthermore, the old distinction between the high culture associated with the upper classes and the mass culture of the working class has crumbled, blurring the boundaries between the classes themselves in perception.[1] Finally, they add, the expanding media industries – television, film, print media and the Internet – and advertising bring a profusion of information and images from across the globe exposing and promoting distant events and ways of life irreducible to extant hierarchies or social divisions. Taken together with the changes in the economy, the detractors continue, all these processes have effaced the old class cultures embedded in occupational communities and famously documented by Hoggart (1957), Dennis et al. (1969) and Jackson (1972) and established choice, variety and ambiguity across the board, with tastes and lifestyles being cast as expressions of individuality rather than as signs of membership of any distinct category. If there are any distinguishable collectivities to speak of in this mass of individuals then these are *status* groups based on, for example, subcultural or 'neo-tribal' symbols (such as those of punk or goth), religion or ethnicity – the last two of which are seen as of increasing importance given global migration and the emergence of multiculturalism as a theoretical and political issue – rather than economic divisions (Pakulski and Waters, 1996; Bennett, 1999).

Finally, the changes in the economic and cultural spheres have, some claim, undergirded and dovetailed with a decline of class politics in the last few decades, evidenced in the incessantly contested psephological fact that manual workers no longer tend to vote predominantly for left-wing parties or white-collar workers and the self-employed for right-wing representatives (Särlvik and Crewe, 1983; Clark and Lipset, 1991). Rising levels of prosperity, occupational shifts and, crucially, the emergence of 'post-materialist' issues and dilemmas through the 1980s – the threat of nuclear disaster, increasing environmental damage, equality for women, ethnic minorities and homosexuals – have, it is argued, ensured that the material issues that once propelled traditional class politics – taxation levels, working conditions and nationalization versus privatization – no longer play a significant part in shaping individuals' political attitudes, activism or ballot box decisions (Inglehart, 1977, 1990; Clark, 2001; see also Mercer, 1990). As a result, the 'forward march of labour' has, in the words of Eric Hobsbawm (1981), been halted and superseded by a flood of 'new social movements' addressed to the post-materialist issues. In reaction, theoreticians of the left 'retreated from class', as Wood (1986) put it, and championed the social movements as sites of struggle capable of inaugurating the better society the working class failed to achieve – shifting from Marxism to post-Marxism (Laclau and Mouffe, 1985; cf. Gorz, 1982; Hall and Jacques, 1989) or at least to a heavily reconstructed version (Habermas, 1987) – while left-leaning political parties in Europe, Australia and the USA embraced centrist tropes of a 'third way' and even the most staunchly progressive reinvented themselves by jettisoning their commitment to the working class and socialism. In the process, and aided by the dominant individualist, free-market rhetoric and policies of the Reagan–Thatcher years, much of which was absorbed into the UK's New Labour and similar administrations (such as the Labor Party in Australia under Hawke and Keating), and the changing concerns of social and political scientists on the left and right, the idea of class ceased to frame political debates over socio-economic inequality and instead discourse on 'social exclusion' or the 'underclass', said to consist of the long-term un(der)employed and other marginalized groups, took centre stage (see Westergaard, 1992; on New Labour see Fairclough, 2000; Levitas, 2005; cf. Boltanski and Chiapello, 2005: 296–314, and Wacquant, 2004a: 108ff, on similar processes in France and the US respectively).

A new antagonist: Reflexivity

Amid these myriad changes, no matter the realm of life and whatever the theoretical vocabulary, one challenging theme has, like the thesis of embourgeoisement forty years earlier, persistently surfaced: whether in the sphere of work, careers and educational pathways, or in the domain of lifestyle pursuits, identities and politics, and whether enforced or enabled, there is, it is claimed, a cosmic proliferation of options and, consequently, a demand for active, individual choice and deliberation no matter what sociological category one belongs to. This has been implied or touched upon briefly by the countless internuncios of new times, from globalization theorists to those still demonstrating a penchant for postmodernism, but none have isolated and elaborated it as clearly and comprehensively as Ulrich Beck, Anthony Giddens, Zygmunt Bauman and Margaret Archer.

This is not to say they share exactly the same conceptual language and causal logic – Beck and Bauman use the term 'individualization', for instance, whereas Giddens prefers to talk of the 'reflexive project of the self' and Archer of the spread of 'autonomous reflexivity' – or even that they necessarily agree with each other, but nevertheless they converge on this common theme enough to be clustered into a distinct intellectual consortium (cf. Warde, 1994; Howard, 2007). Neither is it to declare that they are the first to write in depth on this topic, as if the sprouts of their thought had no roots in the intellectual field. Peter Berger, at whom Beck aims a grateful nod in *Risk Society*, anticipated much of what Giddens and the others would claim when he wrote in the 1970s of the 'individuation' and proliferation of choice brought by modernity

(Berger, 1977: 75ff) and the fact that biographies and identities have become 'design projects' to be worked upon (Berger et al., 1974: 71ff; cf. also Luckmann, 1983: Part 2). Nevertheless, the new theories of reflexivity have been forwarded with the persistence, theoretical depth and timeliness to afford them immense influence. Thus they figure in today's textbooks on sociology and social theory as key interpreters of our times, are hailed by anti-class theorists such as Pakulski and Waters (1996) as significant influences and allies in their assault on class and are cited ceaselessly by class analysts as paradigmatic of the sort of position they are battling against (see, for example, the 2005 special issue of *Sociology* dedicated to class in which Giddens and Beck appear repeatedly). It is, therefore, hard to deny the considerable bearing of these thinkers on the present and the future (and in some cases, paradoxically, the past) of the concept of class.

From heterodoxy to cultural class analysis

Class analysis has not, however, stood still through all this. All of the changes and challenges reported above could be, and indeed have been, countered by class analysts and others on the grounds of their empirical existence, the nature and extent of their consequences for class and the conceptualization of class held to be in decline. To give just one example, the apparent arrival of post-industrialism and consequent diminution of the traditional proletariat has hardly troubled non-Marxists willing to adjust their class schemes to the new scenario and identify a new working class, or new part of the working class, located in lower-level service occupations such as cleaning, fast food service and the like (Esping-Anderson, 1993). Nevertheless, the close of the twentieth century was a turbulent period for class analysis, with a whole array of critics, some of whom once researched and theorized the concept themselves, proclaiming class redundant, dead or dying on the one hand while a faithful knot of defenders vociferously asserted its continued salience on the other (see the debates in Lee and Turner, 1996; Clark and Lipset, 2001; and those between Pakulski and Waters and their critics in the 1996 volume of *Theory and Society*; see also Hindess, 1987; Pahl, 1989, 1993; Goldthorpe and Marshall, 1992; Adonis and Pollard, 1997; Kingston, 2000).

At the same time, faced with the unrelenting calls for class to be abandoned, mounting dissatisfaction with the limitations of the dominant approaches and broader moves within the sociological field, particularly the increased interest in cultural processes (the 'cultural turn'), the field of class analysis began to transform and, in many ways, to revitalize itself – reflected in claims from practitioners that class analysis was 'fragmenting' (Crompton, 1996, 1998) and, later, being 'renewed' (Crompton et al., 2000).[2] Heterodox conceptualizations of class were proffered that challenged the hegemony of the Wright–Goldthorpe couplet by taking their lead not just from Marx or Weber but from figures previously alien to contemporary studies of the concept such as Durkheim and Ricardo, all of which claimed to provide a 'sounder base for class analysis' (Sørenson, 2000), to better capture empirical processes and to more effectively counter the propositions of class critics (see the showcases in Crompton et al., 2000; Wright, 2005). Many of these continued to share features of Wright's and Goldthorpe's perspectives, such as the commitment to primarily quantitative research and economism, but one cluster of writers in this plurality has engineered a significant departure – sometimes bringing previously marginal themes and approaches into the ascendant and sometimes subverting extant divisions altogether – and had a profound influence on class theory and research over the past ten years or so: the so-called cultural class analysts (Savage, 2003: 536).

The core characteristics of this stream of thought – a heterogeneous collection of like-minded researchers rather than a 'school' of any kind who count among their key exemplars Skeggs (1997, 2004), Reay (1998a) and Savage (2000) – are twofold. On the one hand, they recognize to a greater degree the kinds of social changes and challenges documented above and, rather than simply asserting the continued potency of class through advanced statistical tests in the manner of Goldthorpe and his colleagues, endeavour to lay bare, usually (though not exclusively) through qualitative research, the multitude of ways in which class has been reconfigured (Savage, 2003: 536). On the other hand, and in order to follow through on the first aim, the cultural class analysts have sought to distance themselves from the occupation-based and utilitarian perspective of Goldthorpe – who, for his own part, responded to the events of the 1990s by attempting to distance himself from empiricism and deepening both his commitment to a form of Weberianism (testified in his defence of the analytical split between class and status in Chan and Goldthorpe, 2007a) and his rational-action theory of agency in which individuals optimize given situational constraints and their knowledge of them (Goldthorpe, 2000, 2007a, 2007b). This second task has been achieved primarily by detailing the ways in which class is reproduced through *cultural* processes and, even if passed over in silence, manifest in identities, consciousness, dispositions and lived experience (see especially Reay, 1998b; Devine and

Savage, 2000, 2005; Savage et al., 2005c). In pursuing this dual agenda the common theoretical foundation has been neither Weber nor Marx, but the vision of class associated with the late French thinker Pierre Bourdieu, particularly his notions of *capital* (in its economic but also, importantly, cultural and symbolic forms), *social space* (a relational space in which agents are positioned dependent on their possession of different types of capital) and *habitus* (the dispositions formed out of practical engagement with the materially shaped environment shared by those close in social space).[3]

Testing reflexivity

For all their innovation and sensitivity to social change, however, the cultural class analysts, like other class researchers, have failed to respond adequately to the challenge posed by the notion of reflexivity. The result is a perturbing and indefensible void, with some scholars blithely sidestepping it by offering hasty dismissals and cutting themselves off from the main currents of sociological and political debate and others content to nibble at its edges rather than step into its epicentre and effectively assess the intellectual wellbeing of class. Over the following pages, therefore, the theories of reflexivity will be subjected to a direct empirical examination with the intention of discerning with some degree of certainty whether they have the credibility to match their pervasive influence. This will involve two steps: first of all, given their theoretical weaknesses, the reformulation of the core postulations in a more agreeable conceptual idiom, namely – though with a few modifications – that put forward by Bourdieu; and secondly, the subjection of these hypotheses to the trial of research to see if they can, as Popper (2002: 10 *et passim*) would say, 'prove their mettle'. The study, therefore, and somewhat rarely for qualitative research, is primarily 'deductive' in its approach – it is, in other words, 'testing' a set of hypothesized theoretical themes – even if, as with any research act of this kind, there are inevitably inductive moments too – that is, the formulation of new themes and theoretical propositions out of empirical material. Yet the epistemological guide in this venture is not positivism or Popper's critical rationalism (the usual foundations for deductive research, including, implicitly, that of the *Affluent Worker* series), but Bourdieu's postpositivist maxim, inspired by Gaston Bachelard's 'applied rationalism', that the social fact is won (it breaks with lay experience), constructed (it is built into a formalized model) and *confirmed (or, it may be added, confuted) by empirical research* (Bourdieu et al., 1991a: 11).

The central empirical tool in the endeavour to confirm or confute the theories of reflexivity is the interview, just as it was for the *Affluent Worker* research. There are, however, two significant departures from the template study. Firstly, where the Luton team's interviews were mostly survey-like in construction, execution and reportage, the interviews undertaken here explore life histories and subjectivities in considerable depth. This is a pragmatic rather than epistemological decision. Subscribing to Bourdieu's principle of 'methodological polytheism', the method adopted in any one research act is deployed not on the basis of philosophical prejudice – the objective and subjective elements of the social cosmos are, in reality, 'ontologically complicit' (Bourdieu, 1981: 306) and thus open to investigation with the full range of methods – but simply to fit the 'problem at hand' (Bourdieu and Wacquant, 1992: 30). In this case the issue at hand is the veracity of the reflexivity thesis, and while a knowledge of large-scale patterns is crucial for setting the national and international context, and will be marshalled where necessary, the kind of reflexivity described by Beck and the others would be impossible to capture through quantitative techniques deploying closed, fixed-choice questions or tracing 'origins and destinations' because the actual *process* through which the patterning of trajectories or lifestyle practices takes place – constant negotiation of multiple options, or conversely restriction by recognized constraints or tacit pursuance of ingrained possibilities – would remain largely opaque. Instead, qualitative examination of decision-making processes at key junctures past and present, their circumstances and historical antecedents is necessary (cf. Brannen and Nilsen, 2005). The same applies to claims regarding the more subjective element of the theories under examination – the angst, atomization and attribution of individual responsibility theorized by Bauman and Beck in particular or the tacit sense of social distance and the linguistic typifications with which it is articulated, which together reveal classed identity, cannot be rendered in all their analytically vital intricacy in answers to survey questions but instead flow forth best when agents narrate and describe their lives, significant events and relations with others in detail. Largely for this reason the total number of interviews, fifty-five, is considerably less than the hundreds collected by Goldthorpe and his team, but, as argued in the Appendix, this does not diminish the generalizing capacity of the study. Secondly, because embourgeoisement postulated mutations in the working class alone, and specifically their political and industrial orientations, Goldthorpe et al. understandably conducted their interviews primarily with manual workers and fixed their attention on attitudes in the workplace. Reflexivity, however, is a much broader beast, allegedly characterizing whole life paths and the full gamut of values, perceptions and preferences, so the coverage here cannot be restricted to any one section of society or domain of life but must reach across both the occupational spectrum and the life course.

The study is bisected into two parts, with the first section laying the theoretical groundwork upon which the second section, the empirical investigation, rests. Chapter 2 opens the founding conceptual discussion by outlining the theories of reflexivity, grouping them in to two distinct strands, with the ultimate intention of picking out the themes and ideas amenable to analysis. It will then examine the existing critical responses they have garnered from class analysts, showing that, while important points have been ventured, enough has not yet been done to fully confute or confirm the thoughts of Beck and the others as they bear on class. After that, Chapter 3 attempts to reformulate the ideas of Beck and the others into a more agreeable framework in order to generate workable hypotheses. To do this, it outlines the theoretical position on class – that is, what constitutes the historically troublesome concept – adopted herein. Though wary of intellectual fashions and misapplication, the sociological theory of Pierre Bourdieu – grounded in a relational (or 'topological', as Wacquant [2008a] recently put it) ontology and consisting of the concepts of social space, fields, capital and habitus – is logically and empirically compelling and forms the baseline of the investigation. This is not to say, however, that it is without its limitations, particularly in adequately grasping the complexities of social life revealed by in-depth qualitative, life-history research. Individual experiential idiosyncrasy and variation, as well as biographical completeness, escape between the gaps of Bourdieu's concepts, while conscious cogitation and agency remain inadequately theorized. To plug these niggling cracks in an otherwise sound edifice, the link between the French thinker's thought and phenomenology *à la* Alfred Schutz is exploited. To be more precise, Schutz's notion of the *lifeworld*, read in a particular way, and his understanding of the *individual stock of knowledge* are forwarded as useful additions and adjustments to Bourdieu's set of 'thinking tools'. From this position of 'phenomeno-Bourdieusianism',[4] the reflexivity thesis can be critiqued anew but its plausible aspects preserved as testable propositions.

With the conceptual tasks complete, the empirical segment of the book can begin, and as the effects of reflexivity for class are multifaceted, so too must the analysis be. The assessment therefore proceeds on a number of fronts, separating out for analytical purposes two elements of the social world that are, in reality, enmeshed (cf. Bourdieu, 1996a). Chapters 4 and 5 investigate the *structural* dimension of social life as it shapes education and work histories respectively, discerning whether the relational constraints or opportunities of class have, as Beck and the others claim, ceased to play a role in the formation of life paths and been supplanted by reflexive decision-making or whether their pernicious existence continues to dictate trajectories. Subsequently, Chapters 6 and 7 examine whether class has been erased from the *symbolic* realm, probing, first of all, the tacit dimension of lifestyle practices, identities and judgements of others, which Giddens in particular sees as subject to new levels of reflexive construction, and then, secondly, the discursive dimension of class discourse and political attitudes. In all cases the findings are united by a single theme, which will be reiterated and explored further in the Conclusion: the persistence of class relations – the differences, divisions and social distances of objective locations and corresponding dispositions – despite transformations in their substance induced by a social context very different from that of the immediate post-war period.

Part I
Theoretical Preliminaries

2
Reflexivity and its Discontents

The theories of reflexivity may be embedded in abstract and generalized statements about the parameters of contemporary Western societies, and in most cases may be far removed from the practical business of social research, but that does not stop them being, like all models purporting to explain human behaviour, amenable to empirical analysis. Having said that, to proceed properly it is essential to translate the sometimes sketchy thoughts into testable themes by, first of all, isolating the core processes and causal mechanisms postulated and then, subsequently, formulating them into logically coherent conjectures. This chapter will deal primarily with the first step, drawing out as many theses open to empirical investigation as possible. In so doing it will separate out the theories of reflexivity into two categories, each comprising two thinkers, in recognition of the fact that while all four may be united by common ideas and declarations on class analysis, they can, like a beam of white light striking a prism, be refracted into varying hues according to their emphasis, explicitness and causal reasoning. Hence Beck and Bauman are herded under the label of individualization, while the two British theorists are depicted as describing 'late modern reflexivity'.

Once the diverse accounts of reflexivity in 'late', 'reflexive' or 'liquid' modern societies have been suitably adumbrated the chapter will advance to its second task: appraising the various reactions to the theories thus far produced by dedicated class analysts. Again two major strands will be distinguished, this time the quantitative 'falsification' approach of Goldthorpe and his affiliates on the one hand and the 'patchwork' approach of the cultural class analysts on the other. This will allow a further refinement of the themes to be investigated by recognizing already-flagged shortcomings, but it will also underscore the conceptual and methodological *limits* of the responses, which, as a result, fail to nullify the threat to class posed by reflexivity. But first to individualization, and, in particular, to the ideas of Ulrich Beck.

Individualized reflexivity

Ulrich Beck: Class as a 'zombie category'

Beck is perhaps the most obviously anti-class theorist of all those considered here given that, unlike the others, he has been explicitly and persistently proclaiming the waning relevance of class for most of his career – at first with notable qualifications and exceptions, but more radically as his prominence has grown. His position on the demise of the concept is nestled in his general thesis that contemporary Western societies are entering a second, 'reflexive' stage of modernization in which the foundational presuppositions and categories of the first stage, such as scientific reason and the sovereignty of nation states, are being radically undermined as a result of their own cumulative side-effects (Beck, 1992, 1994, 1997, 2009; Beck et al., 2003; Beck and Lau, 2005). Two aspects of this rather broad development, he claims, are particularly consequential for class. The first of these is the changing logic of distribution from wealth to *risk* – a term he is widely regarded as injecting with sociological significance – as rising affluence, the protections of welfare societies and the unintended unleashing of all manner of hazards on an unprecedented scale, such as chemical poisoning, food contamination, nuclear disaster and so on, render the possession of wealth (that is, class position) less significant than proximity to risk in apportioning the primary problems and conflicts fracturing society. Consequently, a whole array of new antagonisms, interests and political movements uniting all victims of risk regardless of background spring up and nudge class struggles and politics into obscurity.

The second, and for our purposes more important, way in which class is supposedly being eradicated from the social landscape in reflexive modernity, however, is via a systematic demolition of the large-group categories of industrial society as the fonts of identities, life situations and inequalities. The key here, Beck argues, is the arrival of *individualization*, the process whereby, under present social conditions, individuals are *disembedded* from 'historically prescribed social forms and commitments' (Beck, 1992: 128), including those of class, and subsequently *re-embedded* in new ways of life in which they 'must produce, stage, and cobble together their biographies themselves' (Beck, 1997: 95). At the root of this, he continues, are the various institutions and welfare state regulations of industrial societies themselves, for these make no provision for the existence of groups. Instead, as he puts it in one of his most quoted phrases, they construct and cater for 'the individual as actor, designer, juggler and stage director of his or her own biography, identity, social networks, commitments and convictions' (Beck, 1997: 95).

Pride of place here belongs to the expanding education system, which, Beck argues, 'recasts and displaces' traditional lifestyles and ways of thinking with 'universalistic' forms of knowledge and language (Beck and Beck-Gernsheim, 2002: 32), furnishes individuals with a capacity for self-reflective knowledge and credentializes them on the basis of individual performance; he also notes the impact of the increased demand for and expectation of mobility and competition in the labour market, prising individuals from their natal locale and kinship networks and forcing them to 'take charge of their own life' (Beck and Beck-Gernsheim, 2002: 32); the democratization of consumption (including access to cars and exotic holidays) with improved standards of living, coupled with a general cultural shift towards 'self-fulfilment' and 'individuality' (Beck, 1998: 39–54); the extension of employment insecurity and instability and, as a consequence, potential poverty across the board under neo-liberalism (Beck, 2000a); the juridification of labour relations; and new urban housing projects breaking up 'ascriptively organized' neighbourhoods (Beck and Beck-Gernsheim, 2002: 35). The result, Beck (1992: 92) asserts, is that class 'now has much less influence on [agents'] actions':

> People with the same income level, or put in the old-fashioned way, within the same 'class', can or even must choose between different lifestyles, subcultures, social ties and identities. From knowing one's 'class' position one can no longer determine one's personal outlook, relations, family position, social and political ideas or identity.
>
> (Beck, 1992: 131)

Not only material constraints and determinations, but – here seemingly targeting the more Bourdieu-inspired – the 'practical knowledge', 'guiding norms' and 'collective habitualizations' furnished by class positions (Beck, 1992: 128; Beck and Beck-Gernsheim, 2002: 6) have dissipated and given way to individual agency, choice and volition in the constitution of life situations (Beck and Willms, 2004: 24). People are thus increasingly forced to construct their *own* biographies and self-identities from the diverse options available and to do so *reflexively* by engaging in 'the processing of contradictory information, dialogue, negotiation, compromise' and 'active management' in the pursuance of 'self-realization' and 'self-determination' or, in short, what is best 'for me' (Beck and Beck-Gernsheim, 2002: 26).[1] Distinctive material and symbolic class differences therefore dissolve 'both in terms of their self understanding and in relation to other groups' (Beck and Beck-Gernsheim, 2002: 39), effacing the identities and solidarity they engendered and, ultimately, removing them from experience altogether (Beck and Beck-Gernsheim, 2002: 36).

Individualization is not, Beck stresses, merely a phenomenon of subjectivity, self-identity or attitude alone, as some writers have argued (for example, Nollman and Strasser, 2007), but a *structural* phenomenon transfiguring objective life situations and biographies. As Beck and Beck-Gernsheim (2002: xxii) put it:

> Individualization can no longer be understood as a mere subjective reality which has to be relativized by and confronted with objective class analysis. Because individualization not only effects the *Überbau* – ideology, false consciousness – but also the economic *Unterbau* of 'real classes'... [it] is becoming *the social structure of second modern society itself.*

Inequality and poverty still exist in reflexive modernity – Beck does not by any means claim society is more egalitarian – but they are no longer differentially distributed between *groups*, as they were in the past, but between *phases in the average work life* (Beck and Willms, 2004: 102; cf. Leisering and Leibfried, 1999). People spiral in and out of economic hardship for an assortment of reasons at different stages of their lives – as university students, as pensioners, after redundancy, following divorce – and this applies as much to the (momentarily) rich as the poor, to managers as much as manual workers. Consequently, individuals should no longer be seen as occupying static positions in a rigid class structure 'handed down from one generation to another' (Beck and Beck-Gernsheim, 2002: 51). Instead they occupy *precarious, ambivalent* positions that are 'subject to cancellation' in a structure conceived not in terms of locations at all, but in terms of *movements* (Beck and Beck-Gernsheim, 2002: 51).

Despite all this, Beck laments, mainstream sociology is hopelessly attuned to the first modernity and its obsolete large-group categories and so continues to discuss and debate the concept of class as if it were alive and well, blindly superimposing it on to a classless society. For this reason he dubs class a 'zombie category': 'the idea lives on even though the reality to which it corresponds is dead' (Beck and Willms, 2004: 51–2; see also Beck and Beck-Gernsheim, 2002: 201–13). But rather than shuffle on solitarily, the living dead tend to stagger forward in packs, and this, Beck argues, spells further trouble for class analysts because it means their flagship concept actually depends on *other zombie categories* for its definition and operationalization. One key example is the idea of a territorially defined *nation state* as a 'container' for the class structure and its conflicts (see Beck, 2000b, 2002). Class concepts, he argues, are 'deeply, intrinsically depend[ent] on the ontology of the nation state' (Beck and Willms, 2004: 104). This is true, he notes, even of Bourdieu's reworking of class: the idea of capital and its exchangeability, after all, 'functions only in a national framework' (Beck and Willms, 2004: 105). Today, however, when individuals from top to bottom of the socio-economic ladder are transnationally mobile 'cosmopolitans' thanks to globalization, the notion of a nation state as an impermeable container is also showing signs of being from beyond the grave. People often have a foot in more than one national framework, each of which positions them in starkly contrasting locations (an Indian taxi driver in Britain who was a professor in his home country, for example), and so to simply look at their 'class' would be to completely misread individuals' self-perceptions, outlooks or choices. In Beck's words: 'The categories of class are simply not differentiated enough to capture such interlocked relationships of border-spanning, multi-perspectival inequality' (Beck and Willms, 2004: 105; cf. Beck, 2007).

Zygmunt Bauman: The individualized society

Given his reputation as a trenchant critic of the vast differentials and endemic misery produced by contemporary capitalist societies, the second proponent of individualization, Zygmunt Bauman, may perhaps seem an unlikely inclusion in today's inventory of anti-class theorists. Yet the view that class plays little part in the maladies of the age of the consumer, first worked out in *Memories of Class* (1982), *Legislators and Interpreters* (1987) and *Freedom* (1988), elaborated in his turn to postmodernism and solidified (so to speak) in his writings on liquid modernity, is clear enough. Indeed, in this latest twist in his thought, Bauman's claims and even his vocabulary become noticeably Beckian, citing the well-known phrases of the German thinker multiple times, though admittedly with important modifications. His core thesis is, in essence, that liquid modernity is characterized by a process of disembedding but – and here is the break with Beck – *without* re-embedding (Bauman, 2001: 41–56, 140–52; cf. 2000: 32–7). The fact is, Bauman says, modernity has always carried disembedding in its train – there is nothing new in that – but in the past this was always promptly followed by a process of re-embedding in which individuals had to actively identify with the new 'communities of reference' (Bauman, 2008: 83) that emerged in the course of social change. In the transition from feudalism to capitalism, for example, individuals were liberated from their ascribed positions in the divinely ordained 'chain of being' – peasant, noble or monarch – and forced to fit into one of the newly emerging social classes, even if, once in place, classes 'tended to become as solid, unalterable and resistant to individual manipulation as the premodern assignment to the estate' (Bauman, 2001: 145), tempered only by the fact that their dispossessions 'added up' and 'congealed' into collective, class interests articulated in communal political action (Bauman, 2001: 46).

In liquid modernity, however, social structures no longer keep their shape for long and jobs for life, once thought to be the norm, have evaporated. The real driving force behind this appears to be the joint rule of unconstrained market forces and individualist consumerism, both of which, to function most efficiently, constantly demand that people change their tastes, habits, identities, affiliations and even occupations, with subcultures and job types springing up and ebbing away incessantly. In any case, individualization has, as a result, mutated: individuals are still disembedded and forced to see their identity as a task rather than a given, but *no longer are there any stable communities of reference waiting to accommodate their self-identification*. Instead, people are permanently disembedded, seeking out, trying on and choosing short-term identities from the immense display of transient options available like shoppers grabbing at a moving clothes rail, all the while feeling incomplete, insecure and unfulfilled.[2] As Bauman (2000: 33–4; cf. 2001: 146) puts it:

> No 'beds' are furnished for 're-embedding', and such beds as might be postulated and pursued prove fragile and often vanish before the work of 're-embedding' is complete. There are rather 'musical chairs' of various sizes and styles as well as of changing numbers and positions, which prompt men and women to be constantly on the move and promise no 'fulfilment', no rest and no satisfaction of 'arriving', of reaching the final destination, where one can disarm, relax, and stop worrying.

It is, in other words, not only 'the individual *placements* in society, but the *places* to which the individuals may gain access and in which they may wish to settle' – lifestyle groups or vocations seem the most obvious examples – that are now 'melting fast' in the neo-liberal climate, and this hits the 'unskilled and skilled, uneducated and educated, work-shy and hard working alike' (Bauman, 2001: 146). The 'problem of identity' for agents is thus no longer 'how to obtain the identities of their choice and how to have them recognized by people around' or 'how to find a place inside a solid frame of social class or category', as it was in the past, 'but *which* identity to choose and how to keep alert and vigilant so that *another* choice can be made in case the previously chosen identity is withdrawn from the market or stripped of its seductive powers' (Bauman, 2001: 147). The idea of a 'whole life project' has become deeply anachronistic, as social conditions now demand that people posses *flexibility*: 'the capacity to forget fast and promptly dispose of past assets that have turned into liabilities, as well as the ability to change tacks and tracks at short notice and without regret' (Bauman, 2008: 66).[3] Freedom to experiment and self-assert in liquid modernity are thus enhanced for 'ever growing numbers' (Bauman, 2001: 50), but not without a hefty dose of insecurity – apparently once a middle-class phenomenon that has been 'spread to the bulk of modern societies' (Bauman, 2008: 47) – to sour their experience.

This is not all, says Bauman, for there is a crucial element to individualization touched on in Beck's thought that is in need of moving centre stage. This is the idea that problems generated by social organization or, more specifically, by deregulated markets and extraterritorial capital, are increasingly misperceived as personal failings and responsibilities that must be dealt with individually. For example, if people

> stay unemployed, it is because they failed to learn the skills of gaining an interview, or because they did not try hard enough to find a job or because they are, purely and simply, work-shy; if they are not sure about their career prospects and agonize about their future, it is because they are not good enough at winning friends and influencing people and failed to learn and master, as they should have done, the arts of self-expression and impressing the others.
>
> (Bauman, 2001: 47)

Private troubles may appear analogous, but nothing more. No longer are they considered to be connected to social organization or collective interests, like those of class, and in fact the company and advice of others simply serves to reinforce the notion that one's troubles must be fought alone through self-determination (Bauman, 2001: 48; see also Bauman, 1999). When it comes to class in particular, Bauman (2004a: 35) states that capital and labour 'no longer seem to offer a common frame inside which variegated social deprivations and injustices can (let alone are bound to) blend, congeal and solidify into a programme for change'. Instead individuals are conceived as and told to be – he has recently betrayed his Marxist roots and described this as an ideology (Bauman, 2008: 88ff) – autonomous, responsible individuals *de jure*, even if, in reality, they remain far from autonomous individuals *de facto*.

But, just as with Beck's thesis, we must be careful not to see Bauman's claims as confined to the realm of subjective identity alone, even if that is the aspect of his work that has gained most attention and garnered approving citation, but as bearing on individual movements within the ever-changing social structure of society and, by implication, on the power of class as a *structural* force dictating life paths. Now Bauman, again like Beck, is not one to deny the existence of inequality and its injurious effects – for example, between those with freedom to consume and experiment with their identity and the excluded 'flawed consumers' and bearers of unshakeable, stigmatizing identities (Bauman, 1998a, 2004a: 38), those with freedom to move around the globe at will ('tourists') and those who either *have* to move because of the inhospitality of the world ('vagabonds') or who *cannot* move for lack of resources (Bauman, 1998b), or simply in terms of a polarization of wealth, income and life chances (Bauman, 2001: 115) – but there is little room for something as rigid and stable as class structure to act as a significant source of advantage or disadvantage in liquid modern life. Peppered throughout his ruminations, especially his more recent tracts, are brief and subtle statements to this effect: 'structures that limit individual choices' no longer 'keep their shape for long' (2007a: 1); individuals are granted a new freedom to 'annul' and 'disable' the constraints imposed by the past so that 'what one was yesterday will no longer bar the possibility of becoming someone totally different today' (2007b: 104); 'assignment to "waste" becomes everybody's potential prospect' rather than 'a misery confined to a relatively small part of the population' (2005: 32) because endemic flexibility and insecurity in the world of work mean that there is an 'incurable fragility of social positions and sources of living' (2008: 75); 'everyone is potentially redundant or replaceable' and 'every position, however elevated and powerful it may seem now, is in the long run precarious' and its privileges 'fragile and under threat' (2001: 52); class divisions have been 'cancelled' (2005: 101), and so on. Most bluntly, when pressed by Nicholas Gane whether his vision of winners and losers in liquid modernity was itself a theorization of a new, rigid class structure, Bauman retorted that 'the two actual, feared or desired social conditions of (freedom and un-freedom)' are *not 'class-ascribed'* (in Gane, 2004: 34). Instead, he claims, they are

realistic prospects for *each and every* resident of a liquid modern society. None of the currently privileged and enjoyable situations is guaranteed to last, while most of the currently handicapped and resented positions can be in principle renegotiated using the rules of the liquid modern game. There is, accordingly, a mixture of hope and fear in every heart, spread over the while spectrum of the emergent planetary stratification.

(in Gane, 2004: 35)

For this veteran of sociology, then, both the identities *and* the life courses of the whole populace have become erratic, changeable, transient and unconstrained.

Late modern reflexivity

Anthony Giddens: The reflexive project of the self

Individualization does not, however, rule the reflexivity roost. Two further positions describing the onset of this class-busting agential condition exist, both of which deploy a different theoretical terminology from that of Beck and Bauman and focus less on the way in which people are *cast* as individuals. The first of these is Anthony Giddens' reflections on the reflexivity induced by late modernity, first formulated some twenty years ago and then gradually abandoned for the world of politics but still, as Adams' (2008) recent engagement testifies, highly influential. The core concept at the heart of this stance, denoting a fundamental reorientation of self-identity and lifestyle, is the so-called 'reflexive project of the self'. If self-identity is, says Giddens, the self – the 'I' – as reflexively understood by the individual in terms of a particular biographical narrative linking the past (how one has become) and future (where one is going) (Giddens, 1991: 53–4), then, in late modernity, the fundamental determinants of this have undergone 'massive change' (1991: 80). The traditional modes of life, including those revolving around class, upon which narratives of self-identity once hung have, so Giddens claims, been 'evacuated' from social life and replaced by 'a context of multiple choice', not because of welfare state mechanisms or unregulated consumer capitalism, as for Beck and Bauman respectively, but owing to *globalization* – the increased awareness and integration into one's own practices of distant events and ways of life – and chronic *institutional* reflexivity – the constant production of new knowledge and information on all manner of issues, from marriage and relationships to health and diet, and its routine incorporation into social practices (Giddens, 1991: 5; see also 1990, 1994a). Consequently, the self has become a *reflexive project* in which people must actively and continuously choose, sustain and revise their narrative of identity themselves. Each of us, he argues, now lives out

> a biography reflexively organised in terms of flows of social and psychological information about possible ways of life. Modernity is a post-traditional order, in which the question, 'How shall I live?' has to be answered in day-to-day decisions about how to behave, what to wear and what to eat – and many other things – as well as interpreted within the temporal unfolding of self-identity.
>
> (Giddens, 1991: 14)

This is usually construed as a positive development by Giddens, being frequently described with the Maslowian therapy-speak vocabulary of self-actualization, self-realization, self-exploration and self-mastery, as it is said to enable hitherto unthinkable measures of autonomy, 'freedom of action' (Giddens, 2002: 47) and 'control [over] one's own life circumstances' (Giddens, 1991: 202), even if, mirroring the anxieties and uncertainties pinpointed by the individualization theorists, it also introduces the new burden of having to constantly reconstruct an 'inherently fragile' narrative of self-identity (Giddens, 1991: 185–6) and fosters a heightened likelihood of shame over the adequacy of one's identity and the inability to match up to one's 'ideal self'.

But this is not all, says the New Labour Lord, for two important consequences follow from this new reflexivity of self-identity. The first of these is an emergence of *lifestyles* in the place of tradition or collective culture. Whereas the latter are handed down from genera-tion to generation and accepted unquestioningly by their possessors, the former, enabled by the 'plurality of possible options' as to how to lead life and the removal of any authoritative guidelines on the matter, signify individual choice and adoption by active agents (Giddens, 1991: 81). As Giddens (1991: 81) puts it, 'we not only fol-low lifestyles, but in an important sense we are forced to do so – we have no choice but to choose'. A lifestyle, he elaborates, is 'a more or less integrated set of practices which an individual embraces [to] give material form to a particular narrative of self-identity' (Giddens, 1991: 81). It forms a cluster of routines, habits and orientations with an overall unity, which 'connects options in a more or less ordered fashion' and removes some courses of action from consideration as 'out of character' (Giddens, 1991: 81–2). Once chosen, however, lifestyles are far from immutable; instead, we are told, they are 'reflexively open to change in light of the mobile nature of self-identity' (Giddens, 1991: 81).

The second consequence, complementing the burgeoning of lifestyle options, is a growing propensity for *life-planning*. With tradition and the sense of fate it imparted now gone, individuals, much like the institutions of late modernity, must 'colonise the future' by integrating projected future actions and events with their past narrative of self-identity in a coherent unity. This is not to say that plans unfold neatly or without disruption, of course – all kinds of more or less contingent events can befall the individual at any time and force them to make high-consequence decisions for the direction of their lives. Such occasions Giddens dubs 'fateful moments', and they include, for example, the decision to change jobs or shift career altogether.[4]

But does Giddens not concede that lifestyle choices are far from equally open to all strata of society, that they may in fact be dependent on the life chances and socio-economic circumstances of particular groups – including occupational groups (Giddens, 1991: 82) and classes (Giddens, 1997)? And does he not, in his influential textbook at least (Giddens, 2006), give the impression that detractors of class have over-stated their case? Well, yes, this is, thanks to Giddens' penchant for ambivalence, all true, but it has to be balanced against the logical corollaries loaded in many of his other claims. Take, for example, the fact that even work in the post-traditional society is now said to be 'by no means completely separate from the arena of plural choices', for 'choice of work and work milieu form a basic element of lifestyle orientations' (Giddens, 1991: 82), or his claims that, though the kinds of constraints and opportunities associated with class still exist, they have, in reality, little influence on the actual social behaviour of individuals (Giddens, 1995: xv; cf. 2007: 65). They are, after all, 'thoroughly permeated by the influence of "biographical decision-making"' (Giddens, 1994b: 188) based on the reflexive 'filtering [of] all sorts of information relevant to [one's] life situations' (Giddens, 1994c: 6) and, in any case, have only a 'refracted' and momentary impact on life chances given the explosion of mobility and unemployment at all levels (Giddens, 1994c: 143–4). Or take his statement that lifestyle choice and life-planning are 'more or less universal': even the most deprived sections of society, he says, can and do, indeed *must*, make self-identity a reflexive project and indulge in the 'creative construction of lifestyle', often 'through the resistances of ghetto life as well as through the direct elaboration of distinctive cultural styles and modes of activity' (Giddens, 1991: 85–6). Accordingly, lifestyles should no longer be considered merely the '"results" of class differences in the realm of production' – if anything, he adds with faint Weberian tones, the chosen lifestyles *themselves* are becoming the structuring features of stratification and social differentiation in late modernity (Giddens, 1991: 82, 228; cf. 1994a: 76; 1994c: 143). A problematic statement for sure, but one that nevertheless underlines the general thrust of his argument: neither self-identity nor social action are tethered any longer to class position.

The consequences of all this, for Giddens, extend beyond the realm of consumption into action and debate within the field of *politics* as well. This is because the emancipatory agenda of class politics hinged upon the equalization of life chances has, so he claims, necessarily receded in importance as *life politics*, or the 'politics of lifestyle' and self-actualization (Giddens, 1991: 214), has blossomed. Addressed to the existential questions and issues brought to the fore by the increasing reflexivity of the social order – such as concerns over ecological risks and the environment more generally, the advance of techno-economic reason and the rights of the individual over their body – the issues included within the remit of the latter simply do not fit within the traditional parameters of emancipation, or even the age-old framework of left and right in politics. Their emergence is premised on a degree of liberation from domination, of course, but they cannot be considered, says Giddens (1994c: 91), merely a phenomenon of the more affluent – to think this would be a 'basic error', for '[s]ome of the poorest groups today (and not only in the developed societies) come against problems of detraditionalization most sharply'. Neither is it the case that emancipatory politics has lost its relevance altogether, but nowadays emancipation often goes hand-in-hand with lifestyle changes and, thus, life politics anyway (Giddens, 1991: 228ff).

Giddens has gone on to build on some of these ideas in outlining his well-known and much-maligned 'Third Way' political programme (Giddens, 1998, 2000). However, since the mid-1990s, with the assumption of directorship at the London School of Economics and the total turn to politics, any rigorous sociological reasoning has been sacrificed for accessibility and so, in these works, instead of talk of late modernity and reflexivity, we get a comparatively trite avowal of the importance of globalization and the 'knowledge society'. There were, however, bolder statements on the mutating class structure of society to the effect that 'the industrial working class is almost ceasing to exist' (Giddens and Hutton, 2001: 5) – the irony being that 'Marx thought the working class would bury capitalism, but as it turned out, capitalism has buried the working class' (Giddens and Hutton, 2001: 22) – and being replaced with routine service or 'Big Mac' workers or, along with the rest of society, the increasingly prevalent 'wired', 'symbolic' or 'Apple Mac' workers employed in the info-tech sector or working with computers (Giddens, 1994c, 2000, 2001, 2007). Just to confuse matters further, though, in his most recent works – still addressed to policy makers and a broad politically interested public rather than sociologists – Giddens (2007) has started to use the term 'class' more frequently again, but in a remarkably superficial, nominalist, atheoretical and unsociological way to refer to arbitrary divisions of occupations that can scarcely be taken seriously. Nevertheless, the emphasis on life politics and its grounding in the prominence of post-materialist issues remains and he continues to assert that classes generally have little bearing on lifestyle practices.

Margaret Archer: Proliferating autonomous and meta-reflexivity

We come, finally, to the most recent antagonist: Margaret Archer, doyenne of critical realism and mother of morphogenesis. Interestingly, her perspective, worked out over the last few years as an elaborate offshoot from an ongoing research programme into the nature of 'internal conversations' (Archer, 2003), is actually premised on explicit criticism of Beck, Bauman and Giddens for overplaying the decline of stratification (Archer, 2007a: 55), but Archer's conclusions are nevertheless conspicuously similar to theirs. True, her strikes at class are perhaps rather more covert and circuitous, though she has her moments, but she nevertheless shares enough in the way of thematic propositions and arguments with Beck and the others to be the fourth figure in a quartet and does, without a doubt, logically suggest a withering of the conduct-shaping powers of class in the current aeon.

Like those of Giddens, Archer's musings on the changing social scene are embedded in an elaborate theorization of the self, in particular on how it mediates between that age-old couplet, social structure and human action. For her, reflexivity, by which she means the capacity to reflect on and transform oneself, is not a new-found, socially induced propensity, but a fundamental capacity of *all* human beings. In fact it is through this reflexivity, which takes the form of an internal conversation – the mental 'self-talk' that we conduct in our heads using language, images, sensations and symbols (2007a: 67) – that humans have causal efficacy in the social world. This is because, says Archer, within these internal conversations, unfolding as dialogues between our past, present and possible future selves that generate non-socially mediated experimentation, creativity and innovation (2007a: 70), each of us considers, deliberates, reorders and confirms our unique 'ultimate concerns' – our priorities in terms of physical well-being, practical competencies and social self-worth – which in turn act as the wellsprings for our projects. Social structures, conceived according to the principles of morphogenesis as sets of external constraints and enablements clearly separable from agents, then enter into the equation by shaping the situations and circumstances people find themselves in and attaching 'opportunity costs' to certain projects that then have to be mulled in the mind (2003: 122; 2007a: 136), but commitment to concerns can ultimately lead to a disregard for these costs and a surmounting of the structural constraints laid in one's way. This is, therefore, a self-confessed voluntaristic theory of action (2003: 122), with reasons being considered causes of action (2007a: 97).

So why might commitment to ultimate concerns surmount constraints in some instances but not in others? The explanatory principle here, says Archer, lies in the fact that people are not all reflexive in the same way. Instead, several modes of reflexivity intermingle in individuals, though with one always emerging as the most characteristic. First of all, there are the 'communicative reflexives': those who require that their internal conversations be completed in external dialogue with others, namely, people they know and trust (or 'similars and familiars' as Archer calls them), before a course of action is undertaken – in other words, they prefer to decide matters through conversation with significant others rather than themselves. Then there are the 'autonomous reflexives' – those who conduct sustained and self-contained internal conversations without input from others and build projects thereon – and the meta-reflexives – those who are reflective on and critical of their reflections and critical of society as well. Finally, there are 'fractured reflexives' whose internal conversations engender distress and disorientation more than anything else. These forms of internal conversation, though not correlated with 'class' (sometimes put in quote marks) or any other demographic indictor, save perhaps education (the more educated are more likely to be more autonomous or meta-reflexive) (2007a: 97), are not randomly distributed but linked to particular social conditions. It seems, for example, that communicative reflexivity is encouraged by, and in turn perpetuates, 'contextual continuity' – where domains of experience and action remain socially, culturally and geographical homogeneous from birth – and hence the existence of similars and familiars, given that this means the internal conversation is required to do less work and can be truncated. The result is that communicative reflexives tend to stick to what they know, evade opportunities for change and remain socially immobile. Archer (2007a: 46–8) sees this as essentially what Bourdieu posits with the supposedly outdated idea that conditions of existence produce a class habitus. Autonomous reflexives, on the other hand, are forged in a context of contextual *dis*continuity, where experience is radically altered and heterogeneous over time, and tend to be more strategic and, ultimately, upwardly mobile as a result. The meta-reflexives, finally, are the product of contextual incongruence and unsettlement, and are consequently somewhat subversive, willing to forgo lucrative opportunities in pursuit of their ideals, and have volatile movements in terms of social mobility. Unlike Beck and the rest, Archer attempts to demonstrate all this with original research, parading an array of interviewed individuals as exemplars of each mode of reflexivity, its roots and its results.

Now these social conditions, claims the matriarch of the critical realist movement, are linked to epochal shifts. In traditional societies, and even in early modern societies, routinization and custom held sway and hence contextual continuity, communicative reflexivity, the class habitus *à la* Bourdieu and immobility were widespread. With the sweeping changes of 'high modernity', however – increased global mobility and travel, expanded higher education, the arrival of the Internet and mass mobile phone ownership – opportunities abound, routinization and tradition have fallen by the wayside, collectivities such as class, and the existence of collective conditions of existence producing a habitus, have declined and people are now experiencing contextual discontinuity and incongruity on an unprecedented scale (2007a: 37ff, 317ff; 2007b: 44ff). The logical consequence is that communicative reflexivity, and immobility, is disappearing while autonomous reflexivity and mobility, and meta-reflexivity and volatility, proliferate, with the overall outcome that 'modernity's long obsession with social stratification and social mobility' is finally 'rejected' (2007a: 324). There may be a whole different vocabulary and theoretical backdrop, and it may revolve around *types* of reflexivity rather than reflexivity *per se*, but at the end of the day Archer ends up with a thesis not that dissimilar to Beck or Bauman on disembedding or Giddens on detraditionalization and globalization, only with a resoundingly positive spin and implied vaunting of global capitalism that surpasses even Giddens.

Wrestling with reflexivity

These, then, are the four major recipes for reflexivity. They toss variegated conceptual ingredients into the theoretical pot – disembedding, personal responsibilization, detraditionalization, contextual discontinuity or incongruity – and identify diverse causes – welfare state institutions, the demands of consumer capitalism or globalization – but in each instance the final outcome appears remarkably similar: increased choice, deliberated decision-making, variation and freedom in terms of life paths and self-conception as a result of social change.

Partly for this reason, pointing out specific theoretical weakness – from Beck's astonishingly lax understanding of class and Giddens' poor treatment of motivation to Bauman's numerous contradictions (see Atkinson, 2007a, 2007b, 2007c, 2008) – provides only a partial response, as a guiding set of *general* ideas common to the different perspectives obviously lives beyond any specific phraseology. Insofar as these broad ideas have successfully captured the mood, furthermore, Beck and the others can, as Savage (2000) has observed, appeal to 'common sense' to push their message through and shrug off this or that analytical deficiency (see, for example, the rather indifferent admission of ambiguity in Beck, 2007).

Yet it is precisely the task of the sociologist, equipped with the tools of science, to pierce 'common sense' and unveil whether the prevailing *Zeitgeist* is built on the shifting sands of rhetoric or the firm foundations of observed reality. Unfortunately, however, though various researchers have engaged with the theories of reflexivity in multiple ways – save Archer's, which is still relatively fresh off the press – the *inadequacies* of these responses mean that this charge has as yet been only partly fulfilled. This can be demonstrated by delineating the two foremost reactions to reflexivity from those within the international field of class analysis, starting first of all with John Goldthorpe's Popperian assault.

Failed falsification

Having earlier doggedly defended class from previous detractors in a renowned empirically-led paper (Goldthorpe and Marshall, 1992), the nature of Goldthorpe's (2002; also 2007a: 91–116) stinging criticism of Beck and the others comes as little surprise. To begin with, he claims, the theories of reflexivity so confidently espoused have about as much credibility as they do empirical backing – in other words, virtually none. Here Goldthorpe joins the multitude of scholars accusing Beck and the others of propounding a 'data free' brand of grand theory hopelessly detached from the world of empirical reality and relying on the force of argument and the appeal to popular ideas to hammer home their view (Marshall, 1997: 16; Savage, 2000: 105; cf. Bradley et al., 2000; Skeggs, 2004: 53; Brannen and Nilsen, 2005; Mythen, 2005a; Fevre, 2007), but Goldthorpe, a dedicated empirical researcher long critical of what he calls 'socio-cultural punditry' (1991: 419), is undoubtedly the most systematic and scathing participant in this line of attack. Denouncing the claims of

Beck and similar theorists as 'without foundation' (2002: 11), 'fanciful' (12), 'more a matter of assertion than demonstration' (14) and ignorant of countervailing evidence (17), he reprimands their use of social scientific evidence for being 'at best patchy and selective and at worst nonexistent' (22), consisting largely of deferential references to the abstract concepts of like-minded theorists or a few select empirical studies, and claims that, as a result, they can 'scarcely be taken seriously' (11).

Goldthorpe and the others certainly have a point. Bauman, as perceptive as he may seem to some, forwards his claims on the basis of virtually no empirical evidence and sticks instead to the writings of others and, at best, anecdotes, while Giddens rests his theory of reflexive self-formation almost exclusively on the content of contemporary self-help books – a source that can only ever furnish a very particular perspective on the contours of everyday life in late modernity. Beck, on the other hand, is a slightly more complex case. While he might not sully his hands in the research process to the same extent as Goldthorpe he does occasionally use statistics to buttress some of his points (see especially Beck, 2000a), he has made reference to qualitative research (Beck and Beck-Gernsheim, 2002: 32) and, in places, he displays a sensitivity to the translation of his ideas into viable research programmes (see, for instance, Beck et al., 2003: 19ff and see also 29n1; Beck and Willms, 2004: 185–7; Beck and Lau, 2005). But then again, the references to statistics and qualitative research are few and far between, rudimentary to say the least and often superficially deployed, while, perhaps most surprisingly of all, in one focal essay he mobilizes fictional characters in novels as his primary font of evidence (Beck and Beck-Gernsheim, 2002: 1–21). An imposing edifice of bold assertions has, it seems, been built on flimsy foundations.

But there is more, Goldthorpe (2002: 20) contends, for when the relevant literature *is* surveyed it quickly becomes clear that the kind of processes Giddens and the others posit as widespread, 'insofar as they are in evidence at all, turn out to be far less dramatic, far more limited and also far more cross-nationally variable than the authors in question would suppose'. It is therefore no surprise that they cite little in the way of evidence, Goldthorpe (2002: 13) jibes, since little can be found. A similar conclusion is reached by Breen and Rottman (1995: 154–5), colleagues of Goldthorpe who claim allegiance to his definition of classes and its attached rational action theory, in their brief review of the literature around life chances: educational attainment, social mobility and myriad other areas – and here they appeal to Ivan Reid's well-known compendium of class inequalities (latest edition 1998) – remain, contrary to what Beck (who they take as their target) claims, decisively structured by class. Not content with merely reviewing the findings of others, however, Goldthorpe has also conducted his own analyses to test some of the claims of Beck and others as they relate specifically to economic security, stability and prospects, and ultimately, like a good Popperian, refuted their conjectures and asserted the continued significance of class (Goldthorpe and McKnight, 2006). This undoubtedly deals a considerable blow to the reflexivity thesis, but not a fatal one. For one thing, in line with their general research orientation the counter-evidence marshalled by Goldthorpe and his affiliates has almost exclusively been quantitative in nature, and there is a sense in which undertaking or reviewing investigations of that ilk, while useful, offers only a partial and, in many respects, problematic reply to the theorists of reflexivity. Statistics cannot, for instance, actually reveal in any definitive or meaningful way whether or not biographies are steered by increasingly reflexive decision-making *within* the unequal patterns revealed. After all, Beck (at times at least: Beck and Beck-Gernsheim, 2002: 30–1), Bauman and others (such as Furlong and Cartmel, 2007) state explicitly that what they postulate need not necessarily reveal itself in a comprehensive reorientation of patterns of inequality, but only an alteration of the processes through which that inequality is reproduced and the social meaning attached to it. For evidence of that, the claim goes, qualitative research of the kind absent from Goldthorpe's repertoire is required and, indeed, usually confirmatory, with Archer's own research being indicative. This is exacerbated by Goldthorpe's volunteered model of explanation – that agents are acting rationally given the situations they are in and the information available – which is at one and the same time not actually incompatible with the general idea of reflexivity – explicit decision-making based on an increased level and variety of available information and awareness of situations – but incapable of exploring the possibility of a *differential distribution of the propensity towards or ability for* such 'rationality' or reflexivity (see below).[5] With his mode of analysis, then, it is difficult to fully assess whether some of the claims of the reflexivity theorists have any merit or not.[6]

Furthermore, the whole issue of identities (or 'subjectivity' more generally) and lifestyles, vital to the claims of Giddens and the others, exposes the limits of Goldthorpe's method of defending class. As regards identities, for example – an area generally sidelined by Goldthorpe himself or reduced to political proclivities – not only has quantitative research addressed to the claims of Giddens, Beck and Bauman on the salience of class identities, solidarities and perceptions produced rather more ambiguous results than other facets of Goldthorpe's defence – purportedly revealing at least some support for individualization (Roberts et al., 1994; Phillips and Western, 2005; Heath et al., 2007; Nollmann and Strasser, 2007) – but of course there is the question of how adequately quantitative research of the kind favoured by Goldthorpe captures identities as they are formed, lived, experienced and articulated in their full complexity anyway. Any possibility of, for example, tapping the analytically important ambivalence and anguish of individuals who may be caught between individualized modes of existence and class-based understandings, demonstrated by qualitative research (Savage, 2000; Savage et al., 2001; cf. Brannen and Nilsen, 2005), is lost, as is, more generally, the prospect of grasping the deep-seated subjective experiences of difference and similarity rendered in an idiom alien to fixed-choice survey questions that may, contra Beck and the others, continue to pervade quotidian life.

As to lifestyles, it should be noted that Goldthorpe, finally addressing the issue of culture with Tak Wing Chan after many years of declaring it irrelevant to his interests, has recently asserted that lifestyle patterns – readership of newspapers, musical tastes and attendance at the theatre, dance events and the cinema – are, according to his explicitly Weberian conceptualization of stratification, more in line with status divisions – constructed through patterns of differential association – than class (Chan and Goldthorpe, 2004, 2005, 2007a, 2007b, 2007c, 2007d; cf. Scott, 2002). This does not invalidate the salience of class in Chan and Goldthorpe's eyes, given that class and status are separable, nor does it imply that lifestyles are necessarily reflexively constructed – though their rejection of individualization is less forthright and unambiguous than their confutation of Bourdieu's perspective and, furthermore, based on a weak understanding of Giddens, Beck and Bauman, who are hastily reviewed indeed, and a conflation of their perspective with postmodernism – but it does admit that class has a rather feeble bearing on cultural consumption.

Yet this kind of empirical claim rests entirely on the conceptual definition of class in play, and in Goldthorpe's case it is a narrow, one-dimensional, economistic one that severs class from culture and consciousness by definition rather than seeing them as inextricably interconnected. For Bourdieusians, for whom culture and lifestyles are core to the conceptualization of class, this result is, therefore, hardly satisfactory. Neither is Chan and Goldthorpe's supposed rebuttal of Bourdieu's theory, since it is rooted in a poor reading that misunderstands and downplays the role of cultural capital in structuring social space – such that Chan and Goldthorpe, like the critics of class mentioned above, can demonstrate the significance of education on cultural consumption but fail to recognize this as cultural capital and thus proof of class *à la* Bourdieu in action – as well as an unsatisfactory definition of cultural practice that explicitly excludes cultural knowledge and 'private' tastes from consideration and the imposition of a class schema and linear methods of statistical analysis at odds with the assumptions of the Bourdieusian approach (see Wuggenig, 2007, and the response of Chan and Goldthorpe, 2007e). Overall, then, Chan and Goldthorpe's evidence, like Goldthorpe's more generally, remains far from conclusive or incontrovertible.

A permeable patchwork

One might reasonably hope, therefore, that the cultural class analysts have provided a more satisfactory response to the theme of reflexivity. They work with a Bourdieusian model of class, thus countering some of the problems of Chan and Goldthorpe's programme, and have a reputation for being more open to rethinking class in light of social change and for engaging significantly with major strands of social and cultural theory, so surely Beck and the others have received sustained consideration and been either roundly rejected or rethought? Unfortunately things are not that simple, for while the cultural class analysts have provided a multitude of insightful thoughts and themes that bear on individualization and late modern reflexivity, these do not add up to a definitive assessment. This is partly because each engagement is partial, superficial or unsatisfactory in itself, but also because, when considered as a whole, they end up with conclusions that contradict, or at least sit rather uncomfortably with, one another. This yields a gaggle of responses that, when stitched together, form a patchwork so permeable the theories of reflexivity can easily escape through its yawning holes.

This is clear when we inspect the three distinct battlegrounds on which the cultural class analysts have clashed with reflexivity: theoretical logic, educational research and investigations of culture and identity. On the first of these, in line with previous critics (Lash, 1994; Elliott, 2002; Mythen, 2005a, 2005b) and taking heed of Bourdieu's assertion that denials of class are integral to the struggle to represent symbolically the social space of differences (Bourdieu, 1987: 2; 1991a: 133; 1993a: 57; on Giddens in particular see Bourdieu and Wacquant, 2001), the cultural class analysts have conjured a two-step conceptual counter. To begin with, there is an open-minded concession that the reflexive construction of lifestyles and biographies may indeed be a feature of contemporary

Western societies, but only with the critical caveat – explicitly rejected by Beck (Beck and Beck-Gernsheim, 2002: 206) – that it is *unequally distributed*. Reflexivity is, in other words, a property of the privileged that flows from their distance from the demands of material necessity and their opportunities for global travel and playful consumption, with the rest condemned to a life of restricted options and routinized constraints (Skeggs, 2004). But this is not all, for added to this proposition is the controversial assertion that the theories of Beck and Giddens (though not Bauman, who is actually mobilized as a *counter* to the other two on the basis of his depiction of intellectuals in *Legislators and Interpreters* – Savage, 2000: 151; Skeggs, 2004: 54) are little more than grandiose attempts to generalize this experience – free of constraint and inscription by others – to the populace at large, serving the interests of the privileged and reproducing inequality by legitimizing, or even directly influencing, the prevalent neo-liberal climate emphasizing unbridled individualism, 'choice' and responsibility (Skeggs, 2004: 52–4; see also Reay, 1998b; Savage, 2000: 108; cf. Anthias, 1999).

An interesting and entirely plausible hypothesis, certainly, but one that remains based on a perfunctory and sometimes polemical assessment of Giddens and Beck and a lack of engagement with their actual concepts in the detail that their influence suggests they deserve – or with Bauman or Archer at all. It is, furthermore, assumed without any real evidence or theoretical elaboration that individualization and reflexivity do indeed characterize the middle classes or the elite in some way. This allows, probably against their intentions, at least a partial vindication of Beck and the others, but it also leaves several significant conceptual questions unanswered. Is the reflexivity of the middle classes, for example, really a consequence of the specific causal processes theorized by Beck and the others, or merely of the conditions of existence they have long enjoyed? How would the former, and the idea of reflexivity more generally, fit with the notion of habitus?

Answers to these questions do, however, surface in the related debate over the idea of the 'reflexive habitus'. For Bourdieu the habitus, as a set of durable dispositions and schemes of perception etched into individuals through practical engagement in the world and functioning 'below the level of consciousness and language' (Bourdieu, 1984: 466), is generally seen as inimical to the constant reflexive refashioning by individuals envisaged by Giddens and the others, but recently a stream of scholars have steadily begun to dismantle this bar between the habitus and reflexivity to argue that the latter – as the drive to question and alter aspects of one's life and lifestyle – *may itself be a durable, unreflexive disposition* inscribed into agents by the transfigured practicalities of their late/reflexive/liquid modern social milieu (McNay, 1999; Adkins, 2003; Sweetman, 2003; for an overview see Adams, 2006). The most relevant effort in this movement to bridge Beck and co. with Bourdieu – that is, the only one not concerned with reflexivity and the habitus as they relate primarily to gender – is undoubtedly that of Paul Sweetman (2003), who contends that if the habitus is the product of the practical adaptation to the basic experiences of the individual's situation, and, as the reflexivity theorists state, that basic experience is, under conditions of late modernity, one of chronic occupational insecurity and flexibility, evaporation of tradition, endless pressure towards individuality and self-realization promoted by consumer culture and diversification of informational input and choice with globalization and the expansion of the media, then it follows that an ability or propensity to reflexively transform oneself will become an integral feature of the habitus or, as Sweetman likes to put it, 'second nature'.

This is not necessarily a blanket process, he notes, but may affect different groups to varying degrees depending on their differential exposure to the fundamental experiences of late modernity. This, then, could be the point of integration with the cultural class analysts – perhaps, it might be hypothesized, the middle classes are, despite Giddens' and Beck's urge to generalize, disproportionately subject to the pressures and processes productive of the reflexive habitus and thus clearly demarcated from unreflexive others. Yet this is certainly not Sweetman's argument. His emphasis is on the increasing preponderance of the reflexive habitus among all sections of society and his consideration of differences within this pattern is glancing, and of class in particular slight. The tenor of his treatment of Giddens and the others makes it obvious that, far from seeing them as myopic promulgators of minority modes of life complicit with the iniquitous status quo, he considers them to be accurate diagnosticians of today's social context in need only of fine-tuning. However, what Sweetman's position *does* have in common with that of the cultural class analysts is that it collapses the theories of individualization and reflexivity together as if they were one thesis, obfuscating the detail and specific causal logics of each in the process, and remains firmly stuck at the conceptual level, drawing only on scraps of others' research. This means that gelling the two positions to form a comprehensive response to Giddens and the others would require considerable specification and elaboration, but also that whether this is really desirable or necessary – beyond simply trying to achieve theoretical neatness by interlinking seemingly oppositional strands of thought – goes undemonstrated.

Perhaps part of the reason these conceptual sketches remain detached from confirmatory investigation is the fact that, when the cultural class analysts do descend from the lofty heights of theoretical speculation down to the concrete arena of empirical research, a rather more confused and contradictory message emerges. In educational research, for example – fertile soil for a Bourdieusian approach to flourish given the French thinker's own contribution to the field – there have been several interesting if partial findings which, nevertheless, neither cohere with the position above or themselves combine to make a clear picture. According to some, for instance, middle-class parents display an ilk of reflexivity in strategizing over their children's education and development in an era of anxiety and uncertainty over the reproduction of privilege (Ball, 2003; Vincent and Ball, 2007), yet in other research, if reflexivity, risky biographies and individualization are to be found it is among working-class youth advancing through the new world of higher education while their middle class counterparts sail through unreflexively (Reay et al., 2005). Quite apart from the unanswered questions in each case – in Ball's research, whether the reflexivity can be generalized to other domains of life, whether it is really any different in principle from the strategies of the dominant in previous times, or whether working-class parents, let alone the children themselves, display anything similar; for Reay and her team (of which Ball was actually a member), whether the identified reflexivity continues on in later life and, importantly, how the researchers explain why some disadvantaged children go on to university while others do not – these findings, like ill-matched jigsaw pieces refusing to tessellate, do not sit well with the supposition that reflexivity is a widespread disposition of the dominant alone nor with one another.

This conclusion persists if we switch attention from education to the second empirical province of cultural class analysis, culture and identities, where much of the discussion has hinged around the notion of the 'individualization of class', a thesis which, in its most basic guise, asserts that objective class structures persist but subjective identities and perceptions mask this by fixing on individual traits (Roberts et al., 1994; Brannen and Nilsen, 2005; Nollmann and Strasser, 2007).[7] Two contributions to this Baumanesque theme are particularly pertinent here. First of all there is Skeggs and Woods' (2008) claim that contemporary reality television programmes encourage a responsibilization of and reflexivity among the working class, a re-imagining of their symbolic deficits as matters of personal choice which, with the right guidance from a middle-class expert in taste or behaviour and a little self-determination, can be overcome. A revealing reading of contemporary programming, but one that, again apart from its own limits – the fact that it focuses only on female audiences (for a description of the method, see Skeggs et al., 2008), the inability to examine the actual sedimentation of the programmes' exhortations within the minds and actions of the viewers and thus whether they employ reflexivity within not just their lifestyles but their life courses – goes against the earlier theoretical hypothesis as now it is the working class who are subject to reflexivity-inducing experiences, albeit mediated by the discourse of the dominant.

Secondly, there is the rather more elaborate and sustained version of this argument presented by Mike Savage (2000), a central figure in the cultural class analysis movement. The starting point for him is the explicit claim that if class analysis is to be revivified given its numerous challengers and attenuation in the hands of Goldthorpe and company then it must not only make space for the cultural and organizational dimensions of social life – two areas generally absent or deficient in Nuffield-style studies – but also *assimilate individualization* as theorized specifically by Beck and Giddens (Savage, 2000: x–xi). There are, it seems, three interlaced ways in which class and individualization mesh in Savage's account. The first of these takes place on the plane of objective inequalities and social mobility. Here, Savage (2000: chap. 4) argues, class operates not through macro-level constraints and opportunities bearing down on collectives, as Goldthorpe's mobility studies would have it, but through *individual biographical trajectories*. By this he means that education, ability and work–life career movements are the primary factors in shaping social mobility, and that while these seem to the lay population to nullify class differences by prioritizing the achievements of individuals over their ascribed (class) position, as celebrated by the old liberal class theorists, in fact these three processes remain infused by class – educational attainment and 'ability', for example, are unevenly distributed between classes – and thus smuggle it back in unrecognized. Class is, in other words, dissolved into the actions and accomplishments of individual biographies until it is no longer seen as such.

The second way class is individualized – and this is perhaps the best-known aspect of Savage's argument – is through *identities* (Savage, 2000: chap. 5; cf. Savage et al., 2001). Drawing on qualitative interviews with 200 denizens of the Manchester region, Savage outlines several features of contemporary class identities. Firstly, he notes, most people are utterly ambiguous and ambivalent about their class identity and fail to supply an unequivocal self-placement, even though they routinely use class as a 'benchmark' to place and evaluate people and see it as a salient social and political issue. Furthermore, and here we proceed to the real heart of the matter, Savage claims that people use class labels to make sense of their life paths by marking out their *relational differences from others rather than membership of any collectivity*. More specifically, people employ class labels to differentiate themselves from people 'higher up' or 'lower down' and place themselves in the middle as 'ordinary' or 'normal' – either using the label 'working class' because of its assumption of working for a living and its anti-elitist connotations or 'middle class' because this designates somehow being in the middle. Class identity thus invokes not solidarity and commonality with one's peers but difference and individuality – something Savage has more recently shown, in his re-analysis of the *Affluent Worker* transcripts, to be far from a novel phenomenon (Savage, 2005).

The final nexus between class and individualization occurs in the realm of culture, and Savage's (2000: chap. 6) argument on this front is, in a nutshell, that there has been a shift from a working-class culture of individualization, in which manual labour epitomized individuality through its autonomy and independence from the employer while the middle class represented servitude, responsibility and accountability to the company or organization, towards the hegemony of a middle-class culture of individualization in which careerism and occupational progress have become central to work life and viewed as 'life projects' of self-development through individual enterprise and performance. Once again, Savage emphasizes, class becomes invisible but, because the habitus of the different classes shapes the perception of a 'good' career and the embodied dispositions necessary to succeed (Savage, 2000: 142, 146), it continues to exert its potent effects.

Savage's overall argument is complex, insightful and, in many ways, persuasive. Yet as a comprehensive response to the theories put forward by Beck and the others – which, of course, it never claimed to be – it falls short for one basic reason: in all cases, it is never actually determined whether *reflexivity* is in operation or not. While the proposals on social mobility appear to tackle some of the causal processes of individualization as laid out particularly by Beck – thus recognizing the thrust of his argument while rereading it through a lens sensitive to the persistence of class in a transformed state – it does not engage with, and the evidence he surveys would not supply an effective response to, the issue of whether biographies are characterized in some way by increasingly reflexive decision-making processes. Instead individualization is taken to mean that class works through individual instantiations rather than collective mechanisms, a definition that sometimes wavers between an account of how class processes operate today and an ontological prescription for the study of class *per se* (see especially Savage, 2000: 150). Similarly, on the otherwise compelling unravelling of the individualization of class identities, no consideration is given to whether identities are reflexively constructed or not;[8] whether other identities, practices or lifestyle activities hold more significance for people and whether they morph over time with the vagaries of identity-construction; or whether individuals are exposed to amplified choice and variety in terms of how to lead their lives and how to see themselves. Instead, individuals – most of whom are, incidentally, in professional and managerial positions and therefore unrepresentative of the full occupational continuum[9] – are simply asked whether they think class is important and whether they identify themselves with a class. Finally, Savage's sketch of classed work cultures could be read as suggesting that the careerist middle-class culture incorporates something akin to reflexivity, though he then brings in the habitus without much elaboration, but the reality is it makes no reference to reflexivity or the way in which it may be induced by changing *conditions* rather than cultures of employment, namely, the widespread employment insecurity that rose up the sociological agenda after the millennial milestone.

Interestingly, this dearth of engagement with reflexivity has continued in Savage's two subsequent programmes of research centred more on culture as lifestyles and consumption, the first with Gaynor Bagnall and Brian Longhurst (2005a) and the second as part of a team of researchers attempting to map out the space of lifestyles and its structural correspondences (Bennett et al., 2009). In the former there is an ambiguous position at best as some sections of the sample are described as 'unreflexive' *vis-à-vis* other, culturally 'omnivorous' sections, to use Peterson's (1992) term, without explicitly labelling the omnivores reflexive or exploring in any detail the theoretical implications of the term (Savage et al., 2005a: 170); in the latter there is something of an explanatory vacuum as the notion of a unified class habitus is viewed with extreme scepticism but nothing concrete is forwarded in its place or investigated in the qualitative material, and though it is briefly mentioned that the middle classes display a level of reflexivity this is not really taken to be a principle of action as much as an ability to reflect on one's consumption tastes (Bennett et al., 2009: 177). The stance in the second project may be linked to Savage's (2009; Savage and Burrows, 2008) recent latching on to a 'descriptive turn' in social science which, with Latour, Deleuze and Guattari lurking in the background but without any convincing philosophical argument, advocates a misguided move away from explanation – whether appealing to reflexive decision-making or a structurally shaped habitus – in the social sciences.

Conclusion

All this leaves us with myriad partial engagements with the reflexivity thesis. Theorists and researchers of diverse persuasions have tackled the three thinkers from a number of angles, noted their limitations and added some credible qualifications. Yet, as has hopefully been made clear, within all these responses there still remains a lack of both a systematic, concentrated theoretical dissection and critical interrogation of reflexivity as it relates to class and an adequate empirical evaluation. The existing contributions cannot be somehow yoked together to form a definitive solution either, for too many contradictions abound and gaps through which reflexivity may escape left open. What is needed, instead, is a head-on reply that can approximate that offered by the *Affluent Worker* team against the proponents of embourgeoisement in the 1960s, that is to say, one that, combining conceptual logic and original research, can begin to ascertain with some degree of authority the crucial quandary of whether biographical paths and subjective identifications, tastes and affects are, in some way, individualized and subject to processes of reflexive consideration at all levels or whether, alternatively, reflexivity is either a property of a privileged few or absent altogether. Such a reply will be pursued over the following pages, starting first in the next chapter with an elaboration of the conceptual architecture guiding the research and a critical reformulation of the propositions of the reflexivity theses into coherent hypotheses. After all, if we are to effectively pronounce it alive, dead or transformed by social change in some way and avoid sloppy synthesis then we need to establish what class has looked like, would look like and could look like.

3
Conceptualizing Class and Reconceptualizing Reflexivity

The once unflinching stranglehold of Marxism and the Nuffield programme over class analysis is beginning to slacken, allowing room for a flowering of perspectives envisioning the concept in starkly contrasting ways (Wright, 2005). Yet, as already seen, there is one standpoint in this assortment currently enjoying particularly frequent adoption and discussion: that forwarded by Pierre Bourdieu. The appeal to Continental concepts, in large part motivated by the fact that they knit cultural processes into the very definition of class and thus resonate with the so-called 'cultural turn' gripping postmillennium sociology (see especially Devine and Savage, 2005), has no doubt been profitable. *Theoretically* it has laid bare the fallacies of the utilitarian model of agency employed by Goldthorpe and Wright by identifying the practical, pre-reflexive and dispositional nature of action flowing out of differentiated past social experiences and the inextricability of cultural frameworks and resources in the formation of 'choices' (see especially Devine, 1998; Savage, 2000), succeeded in reconnecting the analysis of class with broader trends in social and cultural theory (see, for instance, Skeggs, 2004; Adkins and Skeggs, 2004) and even facilitated reflection on the moral dimension of class, that is, its invidious role in dictating perceptions of self-worth (Sayer, 2002, 2005). *Empirically* it has granted a deeper exploration of the relational sense of identity, difference and similarity articulated by individuals (Savage, 2000; Savage et al., 2001), the experiential content of differing positions in the social order and, in particular, the denigration and dispossession pervading life in the lower sections (Skeggs, 1997; Charlesworth, 2000), the reproduction of inequality through differential possession of certain forms of capital and its manifestation in everyday life (Reay, 1998a; Devine, 2004; cf. Lareau, 2003) and the underlying dispositions and outlooks marking out differences and orienting action in certain locales (Savage et al., 2005b). Yet the cultural class analysts have not adopted Bourdieu's theoretical apparatus wholesale. Instead the practice has generally been to appropriate those aspects best suited to illuminating the particular empirical processes in focus, readily blend them with insights from other theories and indifferently reject or ignore the rest. Such a 'pick and mix' approach undoubtedly yields some constructive consequences, especially in those instances where scholars have applied Bourdieu's thought to topics he himself examined only cursorily (I think, especially, of gender – see Adkins and Skeggs, 2004), but partial and, at times, superficial application has also generated a noticeable measure of theoretical inconsistency, misunderstanding and, occasionally, misguided dismissal. There is a tendency, for example, to hypostatize classes as real entities with substantial properties by labelling them 'working class' and so on, thus contradicting or at least failing to acknowledge the fully relational and constructionist aspects of the formation of collectivities according to Bourdieu (who himself was, admittedly, liable to terminological slippage), to inappropriately use Goldthorpe's class scheme for categorization purposes (for example, Ball, 2003; Bennett et al., 2009) and, in some instances, to reduce the notion of cultural capital to either 'awareness of how systems work' and high expectations and aspirations (for example, Devine, 2004: 69)[1] or a (quickly discredited) disinterested Kantian aesthetic (for example, Bennett et al., 2009), thus losing the broader, and more explanatorily useful, sense of the term as an inculcated familiarity and ease with abstract, valued modes of knowledge (that is, 'intelligence' or 'culturedness').

To circumvent these kinds of errors, which Bourdieu (1993b, 1997b, 1999) himself constantly warned against, and because his vision of the social world bears fundamental insights and advances over alternative conceptual schemes far beyond the fashionable 'cultural' aspect of his work, not just in class theory but in social theory and philosophy of social science more generally, the approach here is quite different. The starting point, and therefore, importantly, the standpoint from which the theories of reflexivity will be reassessed, is Bourdieu's theory *in toto*. This is not to say that his ideas are without their gaps and glitches. Elsewhere I have argued that his general scheme can be strengthened, especially to cope with the processes unveiled by qualitative research, by extending its phenomenological dimension (Atkinson, 2010), and this applies just as much to his theory of class. So, in what follows, I will not only outline the core pillars of Bourdieu's thought necessary for the subsequent reformulation of reflexivity but insert critical interludes along the way to respond to common charges and introduce elaborations which, while enlarging and enhancing Bourdieu's conceptual toolkit, nevertheless remain consistent with the fundamental epistemological and ontological prescriptions of his perspective.

The Bourdieusian theory of class

Social space

The best entry point to Bourdieu's theory is his substitute for conventional categorical models of the class structure: the 'social space'. Essentially, this is a space in which all agents are plotted according to three axes or dimensions (Bourdieu, 1984: 114). The first of these, running along a vertical axis, is the overall amount of capital that the individual holds, including *economic* capital in the form of wealth, income and property; *cultural* capital, that is, signifiers of cultural competencies; and finally *social* capital, understood as resources based on associations with certain names or titles and personal networks, which can either provide opportunities (such as the 'old boy network') or capital in its other forms by proxy (a friend or family member rich in economic or cultural capital, for example). A fourth capital, symbolic capital, is the form taken by all other capitals when they are (mis)perceived as legitimate.[2]

Capital – most obviously the cultural variety, but there is no reason to restrict it to this form alone – can exist in three states (see especially Bourdieu, 1997a):

(i) *Embodied*, that is, inhering in the mind and body as knowledge, skills, competencies and orientations. In the case of cultural capital this includes valued cultural knowledge (most famously of music, literature and arts, but also of history or politics), linguistic competence and a familiarity with abstraction and theoretical reasoning, labelled 'symbolic mastery' by Bourdieu and Passeron (1990), which can play out in a variety of avenues (including something contemporary like IT: Kapitzke, 2000). Social capital can be embodied too, however, when interaction with others effects dispositions (see below) such as the perception of what is possible, and thus subsumes much that is attributed to network theory.

(ii) *Objectified*, such as cultural goods and items (artworks, for instance) in the case of cultural capital, available money in the case of economic capital and other people, as sources of 'open doors', in the case of social capital.

(iii) *Institutionalized*, as in the case of educational qualifications for cultural capital. Given his comments elsewhere, it seems institutionalized capital is understood as a sub-type of objectified capital for Bourdieu (1984: 110).

Secondly, and importantly, agents are differentiated along a horizontal axis according to the composition of their capital – in other words, whether it is predominantly economic (as is the case for large industrialists) or cultural (such as for higher education teachers). This serves to subsume cultural differences usually analysed under the separate heading of status into the definition of class, increasing the explanatory power of the concept while, as Sayer (2005: 77) notes, bringing it closer to lay experiences of class which tend not to differentiate it from lifestyles and education. It also introduces the phenomenon, otherwise smothered by one-dimensional class schemes, of horizontal or 'transverse' mobility based on the conversion of capitals – for example, through monetary investment in education producing cultural capital – alongside vertical mobility based on accumulation (Bourdieu, 1984: 131).

The third dimension is actually time, or trajectory, which encompasses the movement of individuals and clusters of individuals through social space as their volume and composition of capital evolves. Far from being the deterministic theorist of unrelenting reproduction, as lazy readings of Bourdieu tend to suppose, occupational change, levels of mobility and dynamism are thus built into the very definition of class, surpassing the static categorizations of existing occupational groups produced by Wright and Goldthorpe.

Two points need to be made before moving on. First of all, and most generally, the idea of social space yields a *relational* view of class in which each position derives its meaning from its relations to others – distance, proximity, above, below, between and so on – within the totality, with these distances and relations translating into real *social* distances and relations. This quasi-structuralist approach, which owes as much to developments in the physical sciences, documented by Cassirer (1923), as to Levi-Strauss, is opposed principally to what Bourdieu calls a 'substantialist' view of class in which the meaning of each position is rooted in the substantial properties (practices, behaviours and attitudes) associated with it, but it also stands in opposition to the relationalism commonly claimed for Marx and Weber (see, for example, Wright, 1979) insofar as it is not concrete *relationships* of production or in the market that are constitutive of class, as the two patriarchs of class theory have it, but social positions defined *relative to one other* in terms of volume and composition of capital. A crucial corollary of this is that the axes defining social space are continuous, or gradational, meaning that, unlike for Marx and his followers or Goldthorpe, there are no hard and fast boundaries between classes, though since agents tend to form clusters or 'clouds' in the different regions of social space they can be separated out as classes for analytical purposes. Thus sidestepped are the stale debates that raged for years between neo-Marxists and others over the 'real' boundaries of social classes.

Secondly, we can put paid to concerns registered by several critics over the role of the economic dimension in Bourdieusian class analysis. Crompton and Scott (2005), for example, warn that Bourdieu-inspired studies of class threaten to underplay the economic inequality at the heart of alternative conceptualizations of class; Bottero (2004), attempting to reassert the importance of the Cambridge school's 'interaction distance' approach, declares that the focus on relational difference in social space really amounts to 'individualized hierarchical differentiation' and stratification rather than class; while critics of class more generally reject the Bourdieusian approach for its 'definitional inflation' of a term that should be restricted to the economic sphere alone (Pakulski and Waters, 1996; Kingston, 2000). Of course, all of these not only fail to appreciate the centrality of the material dimension in Bourdieu's scheme – economic capital plays a crucial role in structuring social space, while Bourdieu (1997a: 54) also stressed that cultural capital, since it requires distance from necessity to be accrued, ultimately has its roots in economic capital – but they also imply something of a *conservative* approach, weighted by the dead hand of tradition, in which the concept of class should be reserved for economic processes alone simply because by and large it has been hitherto. As for the critics of class, sticking to a narrow conceptualization is necessary if their argument is not to fall to pieces from one page to the next, as at least one of them then goes on to document the importance of education – the institutionalized form of cultural capital *par excellence* – in shaping attitudes and lifestyles (Kingston, 2000).

Conditions of existence and habitus

The social space, according to Bourdieu, is constructed in such a way as to reveal the maximum differences and similarities between people. This is because those in neighbouring positions within it, by virtue of their capital possession, share similar 'conditions of existence' and conditionings which, in turn, produce within them similar habitus, that is, a complex of durable cognitive and corporeal dispositions, propensities and schemes of perception and appreciation that manifest themselves in tastes and lifestyles. By 'conditions of existence' Bourdieu generally means the agent's relative distance from material necessity and the experiences this generates, with those in the upper regions of social space, possessing plentiful stocks of capital, being subject to an overall distance from necessity while those in the lower sections, holding less capital, are somewhat closer to its demands and urgencies. Through the practical adaptation to frequently experienced situations the objective probabilities of 'access to goods, services and powers' inscribed in these conditions are then – making a 'virtue of necessity' as Bourdieu often likes to say – transformed into the dispositions, schemes of appreciation and tacit expectations of the habitus (Bourdieu, 2000a: 136; 1990a: 60). Thus *Distinction* (1984) documents how on the one hand the dominant class's distance from necessity results in a privileging of 'form over function' and 'manner over matter' not only in the assessment of art but in the choice of food and clothes and in ways of walking and talking, while on the other hand the dominated class's experience of the urgencies associated with less capital inculcates within them a propensity to give primacy to substance and functionality and, therefore, to make the 'choice of the necessary'.

This adjustment takes the form of a subconscious bodily learning process in which the limits and regularities of the world are inscribed into the habitus as a practical evaluation of what goods, practices and aspirations are accessible and reasonable or, as Bourdieu puts it, as a 'feel for the game' and its forthcoming immediacies. In his words, 'we learn bodily' (2000a: 141) – with the body acting as a kind of 'living memory pad' and 'automaton' that 'leads the mind unconsciously along with it' (1990a: 68) – through 'practice rather than discourse' (1977: 87; cf. Wacquant, 2004b).[3] Much of this learning takes place in childhood, where 'familial manifestations of necessity' – 'forms of the division of labour between the sexes', 'household objects', 'modes of consumption', 'parent–child relations', 'domestic morality' and the like (Bourdieu, 1990a: 54; 1977: 78) – feed into the habitus via 'silent censures' (2000a: 167) and implicit and explicit pedagogy, often inculcating their effects through the experience of corporeal suffering and visceral emotion (2000a: 141), as well as through subconscious forms of mimesis and 'sheer familiarization, in which the learner insensibly and unconsciously acquires the principles of an 'art' and an art of living' (1990a: 74). In households rich in cultural capital, for example, the 'bourgeois culture and the bourgeois relation to culture' are acquired 'pre-verbally, by early immersion in a world of cultivated people, practices and objects' including musical parents, instruments and artworks (Bourdieu, 1984: 75).

However, because the habitus is an 'open system of dispositions' constantly subject to new experiences well beyond infancy it is 'endlessly transformed' through a dialectic with its environment (Bourdieu and Wacquant, 1992: 133; Bourdieu, 1990b: 116). On the other hand, agents are statistically bound to encounter similar, reinforcing situations as a result of their objective social conditions of existence (Bourdieu and Wacquant, 1992: 133), and because so much is instilled in childhood and the habitus operates as the individual's lens through which to receive new experiences, it proves to be remarkably durable (Wacquant, 2005). Thus, for example, the habitus acquired in the family underlies the reception of the experiences of schooling, that is, the 'reception and assimilation of the pedagogic message', with the habitus as transformed by schooling going on to frame all subsequent experiences of culture, work and so on (Bourdieu, 1977: 87). On the basis of this process, Bourdieu often describes the habitus as the 'integration of past experiences' (1977: 83) or the 'active presence of the past' in the present (1990a: 56). Yet this does not imply a role for consciousness in the form of memory, for the body

> does not represent what it performs, it does not memorize the past, it *enacts* the past, bringing it back to life. What is 'learnt by body' is not something that one has, like knowledge that can be brandished, but something that one *is*.
>
> (Bourdieu, 1990a: 73)

It should be clear by now that the habitus is not an apparatus of consciousness, but instead functions 'below the level of consciousness and language, beyond the reach of introspective scrutiny or control by the will' (Bourdieu, 1984: 466). The 'unchosen principle of all choices' (Bourdieu, 1990a: 61), it orients action and practices based not on conscious or intentional aims but on the dispositions and inclinations built out of a practical, pre-reflective, corporeal sense of limits and realistic possibilities, leading agents, as captured in the phrase 'that's not for the likes of us', to refuse what they are refused in reality anyway (Bourdieu, 1977: 77). Bourdieu is, however, keen to stress that the habitus is not a mechanistic translation of objective structures into action, but a *generative* and *creative* capacity for thought and action within limits (Bourdieu and Wacquant, 1992: 122). More particularly, he claims, the habitus is predisposed to generate unconscious 'lines of action' or *strategies* aimed at maximizing the agents' profits, whether they be economic or, more importantly, symbolic (Bourdieu, 2000a: 55). In practice this means that agents endeavour, without the intervention of consciousness, to improve or at least maintain their position in social space intra- and inter-generationally through investing and converting their capital in a multitude of ways (Bourdieu, 1984: 125ff; cf. 1996a: 272ff), though in fact optimizing strategies are diffused through all spheres of life – fields, in Bourdieu's terminology – including those, such as the academic or literary field, supposedly governed by disinterestedness (see Bourdieu, 1998a: chap. 4).

Critical interlude I: Dispositions, intention and projection

Now this is, on the whole, pretty persuasive – when it comes to illuminating the realities of practical, taken-for-granted and conditioned human action unearthed by qualitative research and its correspondence with structural locations it advances far beyond the scholastic models of rational action retrofitted to statistical patterns by Goldthorpe and his followers. Yet some of the language is troubling and, the deeper we dig, the more questionable turns of phrase and implications arise. On the one hand, Bourdieu plays into the hands of the plentiful array of commentators who criticize him for swinging too far in the opposite direction to rational action theory, as well as subjectivism and existentialism, by giving short shrift to consciousness, discursive knowledge, intention, deliberation and reflective thought (see, for example, Crossley, 2001; Noble and Watkins, 2003; Sayer, 2005; Elder-Vass, 2007). Thus there are frequent claims that the habitus produces 'spontaneity without consciousness or will' (Bourdieu 1990a: 56; cf. 2000a: 137) or 'intentionless improvisation' (1977: 79), that there is no 'conscious aiming at ends' (1977: 72), that action is guided not by projects and plans, that is, the future, but by the past as embodied in dispositions (1977: 72) or, if anything, by a sense of the upcoming immediacies (or protention) furnished by the 'feel for the game' that drives tennis players to rush the net (1990b: 12; 2000a: chap. 6; 2005: 214), that 'thought objects', reasons and motives should never be treated as the 'determining causes of practices' (1977: 21), and that the underlying practical sense is nothing other than social necessity 'converted into motor schemes and bodily automatisms' (1990a: 69). When they *are* brought in to the picture, furthermore, they are, discounting the occasional offhand equation of the habitus with mind (Bourdieu, 1996b: *passim*), usually done so in two conflicting ways: either as prompted by, and thus subservient to, *bodily* postures in the manner of Proustian memory (see, for example, Bourdieu, 1984: 474; 1990a: 68–9), or else in times of crisis when the agent suddenly steps out of their habitus – even the schemes of perception, apparently – and conducts some form of 'rational' action (see, for example, Bourdieu 1990b: 108; 2000a: 64).

But then again, and especially in his earlier work, sometimes a rather different image emerges that is not only more convincing in its logic but more in accord with the realities encountered in qualitative, life-history research. Structural conditions, for example, furnish not simply bodily skills, habits and know-how but cognitive knowledge and attitudes (Bourdieu and Passeron, 1979: 17), the habitus operates through 'wills and intentions' (Bourdieu, 1981: 308), consists of 'propensities to *think*, feel, and act in determinate ways, which then guide them in their creative responses to the constraints and solicitations of their extant milieu' (Wacquant, 2005: 316, emphasis added), informs 'all *thought* and action (including *thought of* action)' (1977: 18, emphasis added) and, especially in the early research on Algeria (Bourdieu, 1979), gives rise to long-term plans and projects of action. Discursive forms of knowledge, formulated intentions, projection and excogitation are no longer exceptional and separate from the conditioned habitus, in other words, but flow *from* it. Yet it remains true that the precise contents of the habitus and how it generates conscious thought and intention alongside bodily and spontaneous action is never really elaborated in a systematic way, leaving it open to the charge of being an explanatory 'black box' (Boudon, 1998).

To start us along this road I have suggested that the habitus be thought of as akin to what Alfred Schutz dubbed the individual's 'stock of knowledge' (Atkinson, 2010). This way, though space allows only the most basic overview, a number of points can be clarified. First of all, the knowledge accumulated through classed experience that constitutes the habitus is *multilayered*, covering not just the acquired bodily knowledge of enacted motor skills and know-how but also the web of declarative, verbalizable knowledge recallable with explicit memory (Schutz and Luckmann 1973: 105–11), *all* of which ultimately entail the automatic formation of categories, associations, familiarity and expectations (or protention), which, at the neural level, corresponds to the building and strengthening of synaptic connections in the brain (Changeux, 1985; LeDoux, 2002; cf. Bourdieu, 2000a: 136).

Secondly, these different layers of knowledge *combine* to form dispositions (or what Schutz himself called 'syndromes' and 'attitudes') by structuring schemes of perception, so that when a stimulus (whether an object or an event, such as job loss) is sensed – visually, aurally, proprioceptively – it is done so with the full weight of accumulated categorization and protention bringing into play associated evaluations, rationalizations and lines of action. Of course the formation of these dispositions begins from birth and continues slowly and incessantly in the course of recurrent experience, meaning that acquisition, as Bourdieu always stressed, is forgotten and the dispositions themselves – *why* certain associations, thoughts and lines of action are prompted by perception of a stimulus – are usually unreflexive and unverbalized (Schutz and Luckmann 1973: 215–23; 1989: 20–1).

Thirdly, some of Schutz's nomenclature (but certainly not the details of his position) can be commandeered to help us to think through the fact that the actions generated are not *always* 'intentionless', impromptu or corporal but stretch from impulsive bodily action right up to the most consciously deliberated plan for life. Basically, we can distinguish, with Schutz, 'spontaneous conduct' or 'behaviour' from projected action. The first includes action whose intention and execution are temporally concurrent – what Searle (1983), following Anscombe (1957), anoints 'intention in action' – from bodily adjustment, the contingent back-and-forth of social interaction (including conversation) to ingrained routines that unfurl without prior formulation, but also the spontaneous internal flow of thought – reliving memories, imagination, reflection – involving images and internal speech, all of which Schutz described as 'mere thinking'. Needless to say, the various forms of spontaneous activity and subintentional actions can co-exist in any one moment of practical life, with some occupying the 'core' of attentive consciousness while others (especially bodily movement and adjustment) stay in the 'periphery' (Searle, 1992).[4]

Spontaneous activity with an intention-in-action can occur on its own, but, in reality, it often comes as a component part of a larger organizing intention (such as getting up and walking to enact a decision to go shopping) and derives its overall intentional status therefrom (Anscombe, 1957). To make sense of these larger acts, we can add a final layer of action into which spontaneous activity 'shades' in the flow of conduct and thought: that involving a prior intention (Searle, 1983) or, to use phenomenological terms, *projection*. This entails the positing of a future goal to be pursued, which may then be successful, unsuccessful, interrupted, resumed, abandoned, reformulated and so on as subsequent experience intervenes.[5] Projected actions can be of varying distance and lengths in time – from brief activity in a few minutes' time through a trip to the shops in the afternoon, the goal of attending university right up to some kind of (albeit fuzzy) 'life plan' – and involve varying degrees of deliberation – from a quick 'pulse' of thought (characterized by what Searle [1992: 137] calls 'overflow', or the compression of indefinitely extendable mental contents into a 'flash' of thought) to extended rehearsal and mulling of facets and alternative possibilities over time – but their distinguishing feature, lifting them out from 'mere thinking', is a temporal gap between formulation of the intention and execution of the action. The intended act is always, logically, a 'whole' and may involve numerous automatically assumed steps and the mobilization of various spontaneous acts with intentions-in-action, but this does not (as implied by Schutz, 1972: 59f) necessarily translate into the ingredients of the prior intention. Sometimes we may mentally envision (however fleetingly) ourselves *doing* the act at some stage in the first or third person, or likely scenarios, or the setting, or sometimes we may 'feel' ourselves doing it or formulate it in compressed inner speech, or run through a combination thereof. At other times, as Merleau-Ponty (2002: 207) points out, thought, and thus projection, is 'accomplished' in *speech* – that is, expressed through it in real time – meaning that projects can be and often are worked out in conversation.[6] In the course of being realized the goal is often, depending on its time scale, no longer held in mind as such and, indeed, intended courses of action – and we are, at any point, in the midst of undertaking or waiting to begin multiple overlapping projects (cf. Giddens, 1995: 35) – can involve repetition and settle into taken-for-granted routine activity demanding only intentions-in-action.

To stress it once again, *all* forms of thought and action, including rationalizations of and reasons for action, flow from the explicit and implicit associations, categorizations and expectations of the classed habitus as it meets the solicitations of the world. To give a more 'middle-range' example of the genesis of classed projection, corresponding to what Bourdieu (1984: 110) called our objective field of possibles – our set of objectively possible actions and movements in the social space given our capital and associated experiences – is the agent's *subjective field of possibles* – that is, the wider or narrower set or type of positions and actions (such as choice of job, clothes, or whatever) that enter mundane conscious thought as 'reasonable' or 'doable' (or simply 'me') in the practical business of decision-making. It is the bounds of thought, the range of what is thinkable, mulled and projected – often translated into what agents 'want' or 'like' – based on the associated protention of what is actually possible or likely. So there is intention, there is projection, and there is consciousness, but – unlike in models of unsocialized reflexivity or rational choice – they all remain structured by and founded in the complexes of knowledge and perception constituting the habitus. To put it another way, people act consciously without being conscious of the principle of their conscious acts, and to distinguish this from voluntaristic connotations of conscious action, I will talk of 'mundane consciousness'. Furthermore, where these actions can be traced to a deep-seated competitive inclination, instilled via a complex combination of childhood socialization and later experiences, for oneself or one's offspring to 'do well' or 'better', to conserve what they have or to attain 'self worth' as defined by dominant hierarchies (see below), then they may be called strategies, but if action can be demonstrated not to be guided by this inclination then it is inappropriate to superimpose the language of strategy onto it. The existence and character of strategies is thus more of an empirical question than Bourdieu supposed, as is their differential distribution among agents (cf. Lau, 2004: 378).

Symbolic space

The next component of Bourdieu's theory of class, taking us further into the realm of culture now, is the idea that the practices and consumption tastes generated by the different habitus map into a relational space of their own – the 'space of lifestyles' or 'symbolic space' – homologous to the social space. In other words, corresponding to the different sections of social space are different practices, goods and activities which, because of their homologous distribution, function as signifiers of one's position. For example, golf is plotted in the section of symbolic space which, if laid over the top of the social space, corresponds with the section occupied by industrialists and commercial employers, while football is plotted low down in the position homologous with that of manual workers. More generally, practices and goods based on the dominant taste for form, manner and distinction cluster in the upper regions of social space, bisected according to whether it takes the lavish form of those with predominantly economic capital (luxury cars, boats, expensive holidays) or the ascetic form of those with primarily cultural capital (reading, museums, classical music) (Bourdieu, 1984: 283ff), while the practices and goods associated with the dominated class's 'choice of the necessary' gather at the lower end. Between these two extremes the petit-bourgeois display a lifestyle that betrays both their aspiration to the dominant style of life and their insufficient means and dispositions to appreciate it properly (listening to popularized opera, for instance).

This homology between the social space and the symbolic space yields within agents a practical 'class sense' (Bourdieu, 1990a: 140), that is, a relational sense of difference and similarity, of antipathy and sympathy, of 'one's place' and the place of others, and ultimately of distance and proximity in social space, based on the 'reading' of the signifiers of symbolic space borne and performed by bodies (see Bourdieu, 1984: 241–4, 467–7). Two consequences follow from this. First of all, far from interaction being the source of all meaning, as the symbolic interactionists have it, the source of meaning of all interaction lies in the objective system of differences and similarities undergirding it (Bourdieu, 1990b: 127–8). That is to say, the way in which agents behave and act towards one another – for example, being distant, aloof and standoffish, consciously monitoring and correcting one's behaviour in the presence of someone higher in social space, avoidance, friendliness or condescension (in the sense of strategies aimed at *denying* the social distance) – as well as who they are likely to interact and form relationships with, is structured by their 'practical intuition' of the homologies of the spaces (Bourdieu, 1987: 11; cf. McNay, 2008). Secondly, the agent's position in the spaces, its relation to – or rather its difference from – other positions and the agents' vision of their position, coupled with the specific effect of trajectory (Bourdieu, 1984: 111), furnishes them with their social identity (Bourdieu, 1991a: 234).

Ultimately, argues Bourdieu (1984: 175), the differences in symbolic space are organized around 'structures of opposition' homologous to the oppositions of the social space. The central opposition is between the rare or 'distinguished' practices of the dominant and the common and 'vulgar' practices of the dominated, which maps onto the central opposition in social space between those distant from necessity and those in proximity to it. This means that each practice and disposition in symbolic space, like the positions in social space, derives its meaning only from its relations to others – a practice can only be rare in opposition to the common – and that, therefore, the lifestyle of the dominated serves as a kind of 'negative foil' against which the dominant can define themselves (Bourdieu, 1984: 57). In contrast, the dominated for the most part perceive the dominant lifestyle as a *positive* reference point, that is, as legitimate. This is the principle of what Bourdieu (1998a: 9) refers to as 'symbolic violence': 'dominated lifestyles are almost always perceived, even by those who live them, from the destructive and reductive point of view of the dominant aesthetic'. Of course it should be added that the practices are not *intrinsically* distinguished or vulgar, and thus legitimate or not, but are only so when perceived through the principles of division – the ways of dividing up the social world in perception manifest most simply in binary classifications (high/low, fine/coarse, unique/common, strong/weak, and so on) – instilled into all agents' habitus as 'common sense', primarily through schooling, though with the particular spin given by the conditionings of the agent's conditions of existence (see Bourdieu, 1984: 466–84 and Appendix 4). These principles of division are subject to perennial contestation and struggle (the 'symbolic struggle') at both the level of the individual (strategies of self-presentation and manipulation of one's self-image, but also insults) and the level of collectives (the naming and bringing into existence of groups – see below) (Bourdieu, 1990b: 134; 1991a: 239), but because the dominant have more resources, and thus monopoly over the education system, they have more power ('symbolic power') to impose their definition as the legitimate one (Bourdieu, 1990a: 139).

Critical interlude II: Dissonance and individuation

Recently, however, the close homology of social and symbolic space, and thus all the processes that course from it, has been challenged by Bernard Lahire (2003, 2004, 2005, 2008), a sympathetic critic of Bourdieu in France whose influence is apparent on Bennett et al.'s (2009) attempt to take cultural class analysis to the quantitative level and adjudicate once and for all the wheat (of which there is actually very little) and the chaff (the remainder) in Bourdieu's theory (cf. Silva, 2006; Bennett, 2007). The crux of Lahire's criticism is that, when you actually look at people's lives in detail, there is so much deviation from the patterns described, so much internal dissonance, variation and plurality within not just sections of social space but the individuals themselves, that Bourdieu's model has to be rethought. This, in turn, points to the larger criticism that Bourdieu's homogenizing conceptual toolkit of classed conditions of existence and habitus cannot explain individual specificity, or the range of factors that shape the differences and particularities of people otherwise located in similar structural locations.

Lahire's criticism, however, neglects a couple of features of Bourdieu's scheme that can help make sense of individuation. Firstly, it must be stressed that Bourdieu's model of the social space is *fully relational*, meaning that, even if we can demarcate a 'class' in the space and use occupations as rough proxy measures, it is a dispersed cloud of people possessing differing levels of capital and therefore slightly different conditions of existence, experiential probabilities, habitus and practices as a result, even if still within the same 'family' of class habitus (see esp. Bourdieu, 1990a: 60). As full a knowledge as possible of capital possession is necessary to plot agents accurately, therefore, including of the social capital arising from various networks and relationships. This is, secondly, complicated by *trajectory* – whether someone has been vertically or horizontally mobile will differentiate their practices from their immediate neighbours in the other two dimensions of social space. Thirdly, qualitative class theorists must be mindful of the specifying effects of positioning within multiple relational *fields*, or domains of practice, whether as individuals or through institutions. These include not just the fields of art, literature, journalism, or whatever, but also large workforces and the space of educational establishments through which people traverse (Bourdieu, 1996a; Bourdieu et al., 1999: 618).[7]

These three features together accommodate many of the 'heterogeneous social conditions' listed by Lahire (2004: Part IV), such as social mobility, contradictory family or spousal relationships and diverse networks, and therefore render dissonance more comprehensible than supposed.[8] Yet there is still a sense in which Lahire is on to something, even if his own way of putting it is imperfect. For when we descend from the analysis of what Bourdieu (1988: 21ff) called 'epistemic individuals', or analytically constructed agents with only their pertinent properties for the study of the social space or any one field isolated, to the ideographic analysis of 'empirical individuals', or concrete agents in all their socially moulded singularity, in qualitative research focusing on the full depth and complexity of life histories, decision-making processes and visions of the world, it becomes clear that social space, trajectory and field are not enough to fully incorporate all the elements and differences that shape the habitus and, therefore, not just cultural practices but action more generally. These include the specific social and physical make-up of the home terrain (dwelling place, neighbourhood, region), the precise structure and daily life of the family milieu, concrete consociates, biographical events and so on, with a specific subset for what Bourdieu (1987: 4) himself called 'occupational effects', or the concrete effects of working conditions, occupational cultures and workplace dynamics that are irreducible to the hierarchical field relations of large workforces (see further Atkinson, 2009).

Take, for instance, Cannadine's (1998: 171–80) analysis of the origins of Margaret Thatcher's incoherent visions of the social world – in Bourdieusian parlance, the subjective perceptions of her habitus, including her consumerist individualism and (or despite) her acute class sense, which gave rise to her behaviours, tastes and policies. Key here, argues Cannadine, is not just her position in 'the middle' of social space, though this did contribute to the genesis of an antipathy towards the aristocracy and the working class and set the possibilities of her trajectory, but also serving in her father's shop and seeing people as individual consumers regardless of their occupation (a family and occupational effect irreducible to distance from material necessity) and the particularities of her home town, Grantham, with its lack of heavy industry and traditional working class, its celebration of hierarchical relations in civic events and so on.

More generally, though position in social space may indicate an objective and subjective field of possibles – for example, semi-skilled or skilled manual work, or the musical or artistic field, in considering post-school options – a more differentiated account of residual factors such as 'available information, role models and work experience' is necessary for a full explanation of the further narrowing and weighting of options within the fields for any individual (Archer, 2000: 285). Similarly, knowledge of certain events, consociates and so on may more fully illuminate the origins of dissonant cultural practices.

How, then, do we make space in Bourdieu's conceptual apparatus for incorporating specificity *in abstracto* while staying true to the relational logic of the social world? The first step is to recognize that, from a phenomenological point of view, the knowledge and framing dispositions of the habitus are not just the conditioned product of frequently encountered experiences, but the summation of sedimented experience *in toto* (*Erfahrung*), that is, life experience as a whole (Husserl, 1973; Merleau-Ponty, 2002). Unique and contingent events, as perceived through extant schemes of perception, can thus leave a lasting imprint on the knowledge and dispositions of the habitus, and thus on the practices, of the agent. Having said that, however, experience is structured by the fact that each agent occupies a discrete spatial section of the physical universe, with routinized time–space paths tracing their movements between home, work, the neighbourhood and so on with all their implicated interactions, mediated experiences and solicitations for action (see Giddens, 1984: 110–16, 132–9; Thompson, 1995). These different domains and the recurrent concrete objects (tools, clothes, furnishings), people (friends, family, work colleagues) and events that populate them as a product of the intermeshing of lines of social action bound the individual's experiential world – it is 'my world', or what Schutz called the *lifeworld*[9] – and feed into the habitus, its layers of knowledge and the sense of what is normal, familiar, taken for granted and to be expected (or doxa, to use the label that Bourdieu borrowed from Husserl).

Of course the fact remains that, while the full complexity of lifeworld experience sedimenting into the habitus and contextualizing action is unique to the individual, it is structured by their positions in social space, fields, other relations and trajectory, such that the objects, events and people (with their own lifeworlds and habitus) encountered are particular articulations of *types* or *species* that are differentiated relationally. It is, in other words, a case of seeing the general in the particular, or the nomothetic in the ideographic, and this applies equally to both recurrent experiences (such as the 'dead ends', 'closed doors' and 'limited prospects' that manifest the limits and demands of capital – Bourdieu and Wacquant, 1992: 144 n96) but also one-off events (such as certain illnesses), encounters and interventions by others, the occurrence of which is still distributed probabilistically according to structural position (cf. Bourdieu, 1984: 110).

Nonetheless, the lifeworld contains all the detail and extra experience that inflects and falls between the abstract distributional logics of social structures, allowing us to better explain, for example, the differences that lead one agent in the cultural section of the dominant class to pursue, be at ease with and be knowledgeable of music and another art: both agents' past and present lifeworlds involve a cultured upbringing and distance from necessity, but one may well be characterized by musicality (a musical parent, instruments and paraphernalia around the home), the other by all things artistic. Equally, someone from the dominated section of social space may profess an atypical taste for and knowledge of legitimate forms of culture, such as classical music, because of a relatively contingent association of it with a significant biographical event or interaction with a particular consociate.

Class making

So far we have discussed the existence of what Bourdieu calls 'theoretical classes', 'logical classes' or 'classes on paper', that is, classes of agents clustered in the social and symbolic spaces on the basis of similar conditions of existence, habitus and lifestyle practices. What has *not* been revealed, however, is the existence of 'real', practically mobilizable social groups or classes with predefined boundaries, definite criteria of membership or a 'unity of consciousness' or interests (Bourdieu, 1987: 7). In fact, such groups and their boundaries, including those posited by Marx, Wright, Goldthorpe and other class theorists, *never* exist ready made in reality – such an idea is an essentialist or substantialist one – but are instead *symbolic and discursive constructions* constituted in history through the symbolic and political struggles over the legitimate principle of vision and division of social space. In other words, the ideas and labels of 'middle class' and 'working class', 'bourgeoisie' and 'proletariat' are nothing more than *representations* of the divisions and differences of social space, building upon and raising to the discursive level the practical sense of difference and similarity, which, through specific political processes, have come to mobilize individuals in social space and feed into their social identity and sense of belonging. The act of *naming* a class or group is the crucial first step in fostering the belief in its existence, followed by the establishment of organizations, symbols and representatives delegated the task of speaking for and about it, though of course which constructions gain credence depends not only on its concordance with the realties of social space but on the symbolic power of the constructor (Bourdieu, 1987: 8–9; 1991a: 239–51). Once in circulation, representations of social space can have real effects on the distributions within it, especially when they are recognized by the state in law, through, for example, various processes of exclusion and the credentialization of occupations.[10]

Representations are firmly anchored in the social space and the differences it yields and do not, therefore, operate in a 'social void' (Bourdieu, 1998a: 12). Far from theorizing a relativist nominalism, Bourdieu thus elaborates what Wacquant (1989: 173) calls 'constructivist realism', which, first observing the rationalist directive uniting Durkheim and Bachelard to break from and push aside agents' subjective constructions and 'prenotions' and identify the existence of real structures independent of human thought that shape action and form a base for representations – the social space – then recognizes the fact that this reality *is* perceived and constructed by agents, that the objective structures do not 'uniquely determine what social collectives emerge out of it and in what form' (Wacquant, 1991: 60), and that these constructions contribute to 'producing the facticity of the objective world' (Wacquant, 1989: 173; see further Bourdieu et al., 1991a). In other words, the social world is both 'real' in the sense of existing independent of our perception of it and 'constructed' in the sense that 'its mechanisms function only as they are perceived and appreciated by agents through schemes that are socially produced within, and homologous to, objective structures' (Wacquant, 1989: 174).

Proximity in social space by no means 'automatically engenders unity' (Bourdieu, 1998a: 11), guarantees symbolic and discursive articulation or gives rise to mobilized groups – in Wacquant's (1991: 57) words, classes at the symbolic level are 'largely underdetermined at the structural level'. Part of the reason for this is the fact that the 'the relative indeterminacy of the reality which offers itself to perception', the 'plurality of principles of vision and division available at any given moment' as a result of past and present symbolic struggles and the specific twist given by the individual's position in social space as they produce their classifications of the social world to meet the exigencies and experiences of their daily lives (Bourdieu, 1987: 10–11) mean that the divisions of social space can, according to Bourdieu, be perceived, constructed, represented and acted upon by agents in different ways – including in terms of ethnicity, occupation, community and even in terms of the explicit absence of social classes. Nevertheless, when drawing together and mobilizing as a practical group agents distant in social space on the grounds of, for example, nationalism, ethnicity or gender, the social and cultural fissures between them are liable to result in fractures (Bourdieu, 1991a: 232–3; 1998a: 11; cf. 2001: 93).

The separation of theoretical classes from symbolic classes is one of the most refreshing aspects of Bourdieu's perspective and, as Weininger (2005: 116–17) notes, leaves him adequately equipped to recognize processes usually associated with the demise of class – pitching them at the level of symbolic and discursive construction – while continuing to uphold the significance and analytical value of theoretical classes. Thus the decline of traditional working-class identities and solidarities, the fragmentation of their communal heartlands and the demise of socialist politics are all documented at length in the interviews of *The Weight of the World* (Bourdieu et al., 1999), as well as in Charlesworth's (2000) broadly Bourdieusian study of de-industrialized Rotherham, without implying that the distributions and clusterings of social space, the similar conditions of existence and lifestyle practices or the practical sense of proximity and distance have altered significantly.

Critical interlude III: Perception and typification

However, while the bulk of what Bourdieu has to say on this facet of class is both highly illuminating and convincing, laying firm foundations for studying the subjective dimension of class, there is a sense in which his account is undeveloped and in need of some padding out, particularly in terms of the nature and genesis of schemes of perception and constructions. He gives little away, for example, on exactly how perception is moulded by experience other than brief claims that the schemes of the habitus represent internalizations or 'interiorizations' of the structural oppositions of fields or social space (Bourdieu, 1996a: 236), nor says much on how the basic binaries mentioned earlier relate to the complex constructions implicated in class or group making. Once again we can turn to phenomenology to help us out here and, specifically, to Husserl and Schutz on 'typification'.

Earlier it was stated that the knowledge of the habitus operates through categorizations and associations that structure our perception; being more precise now, we can say these take the form of typifications and pairings.[11] Crossley (2001: 132), another thinker with a keen eye on the linkages between Bourdieu and phenomenology, lucidly summarizes the process as follows:

> Typification entails the formation of habitual perceptual schemas which simplify complex perceptual input. In effect the uniqueness and particularity of each new moment of our experience is simplified by being subsumed into a general category or 'type'. Thus, even when we approach objects which, strictly speaking, we have never encountered before, we will see them in terms of the broader type to which they belong. Moreover, newly typed objects are 'paired' with objects of the same type which we have experienced in the past, and properties and qualities attributed to them accordingly.

Far from being conscious phenomena (as implied in the critique of Schutz presented by Ostrow, 1990), typification and pairing occur automatically and 'without our participation', as Husserl (1973: 123) puts it, at the pre-reflexive or 'prepredicative' level – the child who encounters and understands scissors for the first time will thereafter simply perceive scissors *as* scissors at first glance without any 'explicit reproducing, comparing, [or] inferring' (Husserl, 1977: 111). Only when typification is problematic, such as when a perceived object is distant and we are struggling to 'make it out', does it move closer to consciousness. Furthermore, as should be readily apparent, the typifications and pairings comprising a scheme of perception are constituted out of the agent's past experiences and overlap with the varying forms of knowledge and know-how described above (Crossley, 2001: 132), meaning that they too, like the habitus more generally, can be considered structured according to the material and cultural conditions characterizing the agent's lifeworld. Finally, argues Crossley (2001: 133; cf. Schutz and Luckmann, 1973: 233–5), language plays a crucial role as the objectifying vehicle of typifications and pairings, from furnishing individuals with the oppositional adjectives identified by Bourdieu to providing the names and descriptions of the most elaborated symbolic construction of class or another grouping.

With the perceptual dimension of the habitus thus refined we can start to put some new flesh on the bones of Bourdieu's framework. First of all, the 'reading' of the signs of symbolic space giving rise to 'class sense' can be conceived as a prepredicative attribution or recognition of types and pairings. That is to say, the practices and goods of symbolic space are associated, paired and typified with agents in certain sections of social space homologous with their position, such that when a 'sign' of symbolic space is perceived it triggers a prepredicative association of the bearer with a certain position in social space relative to the perceiver (rendered in terms of occupation, wealth, 'intelligence' and so on) and other practices as well as with linguistic descriptors ('posh', 'fancy', 'vulgar') and affective states (loathing, discomfort, fear).[12] The other side to this, of course, is that others' expectations of an agent and their actions based upon them, such as discriminatory practices, shape the agent's lifeworld and build into their habitus, more deeply the more recurrently they are experienced (cf. Jenkins, 1996: 154–70; Crossley, 2001: 150ff).

The typifications constituting schemes of perception are built out of multifarious experiences, often emanating from contradictory sources, and tailored towards practical purposes and considerations (such as insults or descriptions) (Bourdieu, 1987: 10). Bourdieu (1987: 10) is thus right to assert that they 'are never totally coherent or logical in the sense of logic' but instead 'necessarily involve a degree of loose-fitting', fuzziness and incoherence. This is even more the case when those perceived are either from the middle sections of social space where 'the indeterminacy and the fuzziness of the relationship between practices and positions are the greatest' and open to manipulation (Bourdieu, 1987: 12), or from a position distant in social space, outside of the agent's routine lifeworld, and thus grasped with more generalized criteria or by reference to representatives of that sector with which they are familiar (including those from the media) (Bourdieu, 1987: 10). Moreover, though typifications remain structured by the symbolic space as an index of objective statistical associations and the determinate position the perceiving agent occupies within it, because the agent's subjective scheme of perception is constructed out of the practical exigencies and experiences characterizing their lifeworld the signs recognized, the specific meaning they are given and their linguistic objectification are all particular to them. For this reason differences in social space are often grasped at a local level – between concrete individuals or, as in Southerton's (2002) study of class identification in Yate, between housing areas within a conurbation – or in terms that have had the greatest salience in the agent's experience (see Bourdieu, 1987: 10). Having said that, it is important to remember, in line with both Bourdieu and Schutz, that many linguistically objectified typifications and constructions are, as products of past symbolic struggles, appropriated 'ready made' by individuals – from parents during socialization, but more importantly from agencies of symbolic power such as the school, the media and political discourse (the latter of which is mediated by the media: Bourdieu, 1998b; see also Bourdieu et al., 1999: 620) – and applied to their circumstances. One need only think of the contemporary construction of 'chavs', 'metrosexuals' or, a little older now, 'yuppies', all of which are associated with certain clothes, goods and practices and figure in everyday discourse, but even the labels 'working class', 'bourgeois' and, more popular these days, 'middle England' fall within this bracket.

Reflexivity revisited and reformulated

These, then, are the pertinent components of the conceptual system underlying the study. Now we must return to the substantive task at hand and sketch their consequences for the theories of reflexivity and, in particular, the reformulation of the latter into researchable hypotheses. Which ideas, in other words, can be rejected as untenable given the presuppositions outlined above, and which can be reconceived in a new vocabulary with provisos and anticipated counters for empirical test? To answer this question it is useful to introduce the analytical split that structures the remainder of the study: that between, on the one hand, the *objective* or *structural* moment, that is, travels through social space and their propulsion by either reflexivity or classed dispositions, which itself encompasses education and post-educational social trajectories, and on the other hand, the *subjective* or *symbolic* realm, or tastes, schemes of perception and evaluation, systems of typifications and the practical sense of difference and similarity or unbridled individualism and atomization. In each of these domains we can think through the averments of the reflexivity theorists in a phenomeno-Bourdieusian mode and reject, question and reformulate as necessary.

The structural phase

The education system, with its inculcation of specific forms of tradition-busting knowledge and rapidly growing reach, is, for Beck and Archer especially, conceived as one of the foremost factories of reflexivity, endlessly transforming children into fully individualized beings and injecting contextual discontinuity into their everyday experience regardless of conditions of life. The power of inherited capital, whether economic or cultural, is thus reduced to nought and, so says Archer (2007a), the model of educational inequality and reproduction associated with Bourdieu's name rendered an antiquated idea better suited to earlier stages of modernity. But the outright rejection of Bourdieu here is problematic. Neither Beck nor Archer consider that the education system may have divergent effects for different sectors of the population, despite critical concessions that indicate otherwise. Beck, for example, acknowledges that the reflexivity-encouraging effects of education depend on the content and duration of the educational experience (Beck and Beck-Gernsheim, 2002: 32), and if we conjecture, not unreasonably, that these depend on resources available, then it would seem this disembedding mechanism is in fact only equipping *some* – namely those possessing ample stocks of cultural and economic capital – with a form of 'reflexivity'. This could then be cleansed of its voluntaristic renderings, captured in hollow claims that individual volition is somehow more prevalent than before (Beck and Willms, 2004: 24), and recast as an unreflexive *disposition of the dominant* to consider and choose different options with reference to oneself, alter facets of one's biography and follow diverse paths through life in the way Sweetman (2003) suggests. If this is indeed the case – and if Beck's thesis is to hold even this meagre amount of water it has to be shown that the underlying force behind such reflexivity is not simply the distance from necessity the more privileged have long enjoyed – then not only is the German thinker, as Skeggs (2004) claims, depicting as universal a mode of behaviour restricted to the more privileged but, more importantly, class (by Bourdieu's definition) is not dead at all.

As regards Archer, after all her declarations that modes of reflexivity cut across (what she calls) class and render Bourdieu's thought out-dated we come across the unsurprising concession that autonomous and meta-reflexivity – the two more extended and sealed-in varieties of deliberation with oneself involving either strategizing over multiple options or following higher ideals – are *closely associated with higher education levels* or, in other words, cultural capital (2007a: 97 n86, 151 n6), fostering the suspicion that the supposed class-breaking modes of reflexivity are nothing more than the dispositions of those with a privileged education. Pointing to the expansion of university participation rates to try and claim that such an education is no longer bound by class will not avert this interpretation either, at least when it is done in a remarkably undifferentiated way that pays no heed to whether the new entrants are from particular backgrounds (even if first generation) and which types of institutions are attended (Archer, 2007a: 319–20). But we must suspend the hunch that autonomous and meta-reflexivity are simply products of distance from necessity and remember that Archer specifically claims and tries to show them to be adaptations to contextual discontinuity or incongruity – in reality generated and operating in exactly the same way as a disposition of the habitus, which she officially rejects on the basis of a quietly contradicted theoretical commitment to an almost Sartrean unsocialized voluntarism and an incredible reading of the Bourdieusian concept, ignoring Bourdieu's (2000a: 159ff; 2007: 100ff) writing on hystersis and the cleft habitus, which couples it with contextual continuity alone rather than experience in general. Given the nexus between autonomous and meta-reflexivity and cultural capital, this suggests a relationship exists between cultural capital and contextual discontinuity or incongruity. Archer would no doubt see this as a case of the context engendering reflexivity, which then produces cultural capital, but we might instead wonder whether contextual discontinuity and incongruity are themselves encouraged by lifeworlds rich in cultural capital as children in 'educated' homes are disproportionately exposed to different ways of life, cultures, spatial locales, options and ways of thinking from an early age and are therefore simply components of capital transmission.

Much of what the reflexivity theorists say on the restructuring of the post-educational life course is highly questionable too, instantly jarring with the Bourdieusian intuition that trajectories are steered by the relentless inertia of capital. In Bauman's work, we get the image that people are constantly floating around the fluid social structure of society as jobs and livelihoods spring up and vanish with the vagaries of capitalism, and with that the controversial thesis that the social space has effectively ceased to be of any significance in shaping experience, habitus and trajectories. Instead all are, as Sweetman (2003) proposes, subject to more or less the same conditions of existence – chronic instability, pressure to annul one's past and so on – and, as a consequence, develop the same dispositions (anxiety, constant self-refashioning and so on) and suffer the same erratic social movement, or at least fear thereof. With Beck we get the hyperbolic declaration that positions in the social space are characterized by transience, incessant movement and ambivalence, with inequality ultimately being distributed over the phases of an 'average work life' – a phrase that conceals vast differences between occupations (say Giddens' 'Apple Mac' workers compared with 'Big Mac' workers) – rather than between groups. Similarly, Archer (2007a: 61) declares that the actions of 'new cosmopolitans' miraculously owe little to their 'background and socialisation' and thus they cannot be 'Bourdieu's people', while Giddens shares the widespread mindset that mobility and insecurity have removed both the limits and guarantees of capital.

This should all be contrasted with Bourdieu's assertion that movements in social space can only occur with the accumulation (or loss) or conversion of capital, which generally takes a specific amount of *labour* and thus *time* (Bourdieu, 1991a: 232), or the inflation and depreciation in value of particular forms of capital such as educational qualifications. This is not to say that intra-career mobility does not and cannot occur, as some absurd readings of Bourdieu claim, but simply that it requires a quantum of effort absent among those whose capital allows a privileged entry point and a disposition, the genesis of which needs to be traced in the classed lifeworld, to expend that effort. Neither is it to claim that jobs are not lost at higher reaches of social space, but only that those unfortunate enough to be made redundant do not instantly plummet down the hierarchy: they are likely to retain their capital (especially cultural and social capital, perhaps the more important for securing new employment, but also economic capital through generous payouts) and will be, therefore, better equipped not only to job-shift quickly but to do so horizontally (or even upwards) rather than downwards (see Hebson, 2009). They would also in all probability – though the research would need to demonstrate this – retain their dispositions and attitudes or, if the experience of job loss is recurrent, develop new ones – both positive (reflexivity) and negative (such as an inability to grasp the future as in Bourdieu, 2000a: 234) – still distinct, on the whole, from those of the dominated. As to the far-fetched claim that action owes little to background or socialization, this is difficult to accept on logical grounds given that one's habitus and biographical situation at any point is the total summation of embodied and enacted past and that the experiences in early years lay the foundations on which every subsequent experience and action are built. Even an apparently radical departure, induced by some combination of experiences, is launched from this platform and has its direction and distance circumscribed by the field of possibles it opens up to consciousness.

The symbolic phase

On the symbolic dimension – the lifestyles, discourse and politics of class – it is worth beginning with a general point, already broached in the outline of Bourdieu's position on class making, on the frequently asserted weakening of collective class identities and politics and the rise of alternative divisions and issues in their place or the onset of atomization. Far from spelling the end of class *per se*, from a Bourdieusian perspective these processes can be conceived as simply a decline of the *symbolic construction* of 'class' as a frame for articulating the differences of social space and mobilizing agents with the rise of individualist political visions of the social world, particularly in the 1980s, and the increased prominence of alternate constructions of difference such as ethnicity, nationality or 'social exclusion'. The disappearance of some of the symbols associated with discourses of 'class' – certain jobs, communities, ways of life and so on – with changing social conditions could also be an important factor in shaping perceptions and linguistic descriptions of the social world, though it must be made clear that this is only because of the substantialist worldview characteristic of everyday thought in which what makes someone a member of a 'class' is the display of a particular combination of properties and practices with which the label is typified. This must be separated from the relational definition of class of the analyst, where theoretical classes exist so long as differences – relative distances and directions – in social and symbolic space persist and manifest themselves in the *sense* of difference, no matter what the actual symbols homologous with each sector of social space may be or how they are discursively articulated. This applies to politics too: it is not so much the precise content of political debate that matters – whether materialist or post-materialist, for instance – or how it is articulated, but the correspondence of stances on political issues with positions in social space.

With this distinction in mind, it soon becomes clear that much of what the reflexivity theorists argue, while empirically plausible, fails to establish the demise of class. Take, for starters, Beck, who, having clearly failed to break with the substantialist logic of lay thinking, fires off a number of arguments that plainly miss their target. His claim that social classes are 'losing their distinctive traits, both in terms of their self understanding and in relation to other groups' (Beck and Beck-Gernsheim, 2002: 39) and are thus 'no longer experienced' (Beck, 1992: 98; cf. Beck and Beck-Gernsheim, 2002: 36), for instance, can be reduced to little more than the much less threatening proposition that the *particular differences associated with 'class' as a symbolic construction are disappearing*: the old jobs, communities and practices that marked out in people's minds 'the working class' and 'the middle class' as popularly conceived. While strictly speaking an empirical question, the idea that symbolic differences *per se* have disappeared, and with them the sense of difference and similarity based on the prepredicative association of perceived symbols with positions in social space, seems somewhat harder to sustain. For example, while for Beck (1992: 95) the supposed 'democratization' of car ownership and foreign vacations may signal the end of class, in the sense that the working class are obviously defined by their domestic holidays and lack of a car, for a Bourdieusian the differences and relations *within* these 'democratized' practices would signify the continued existence of symbolically differentiated classes – the 'old banger' of the dominated class, or the 'souped-up' machines of younger, usually male members, versus the executive or sports car or four-wheel drive (the 'Chelsea tractor') of the dominant; package beach holidays in cheap destinations such as Spain for the dominated versus independently booked adventurous or 'cultural' holidays in search of 'authenticity' for the dominant, and so on. If such difference exists, whether or not encased in the language of social classes, then theoretical classes exist; conversely, to deny classes on this definition 'means in the final analysis denying the existence of differences and principles of differentiation' (Bourdieu, 1998a: 12).

Bauman fares no better. If disembedding is conceived as a process whereby individuals can no longer identify with fixed groups but are instead compelled to build and rebuild their identities themselves (once again the reflexive habitus suggests itself, but in an identities/lifestyles register), then this need not be too troublesome because, on this view, the decline of capital and labour, or other collective categories for that matter, represents not the withering of objective patterns of inequality but a decline of 'class' as a *frame for interpreting* those 'variegated social deprivations and injustices' produced by the stratification of freedom (Bauman, 2004a: 35), fitting snugly with the ideas already mentioned on the decline of symbolically represented classes. Furthermore, the assertion that individuals are increasingly cast in liquid modernity as either 'flawed consumers' or autonomous and responsible individuals can be reinterpreted as a perfectly credible description of new dominant ways of carving up the social space in perception, especially as propagated in the political sphere by the Thatcher government of the 1980s and largely continued under New Labour. Rethinking one of Bauman's dualisms, it could be hypothesized that *de jure* freedom denotes the widespread perception of social space as composed of autonomous, atomized individuals, with *de facto* freedom referring to the real degree of freedom granted by one's position in social space. However, the same reservation forwarded against Beck has to be entered here for Bauman as well: if symbolic differences homologous to the divisions of social space continue to yield some sense of difference and similarity – as indeed implied in Bauman's description of the vilification of 'flawed consumers', while squeezing out the subtle differences in symbolic space between various areas of social space – then the efficacy of the social space and the theoretical classes it contains in impressing upon visions of the social world remains.

Finally, what about Giddens on reflexive lifestyles? From a phenomeno-Bourdieusian perspective, of course, the habitus formed out of the experiences of the agent's materially and culturally structured lifeworld is the wellspring of action and principle of lifestyles, and because the structuring of agents' lifeworlds depends on their position in social space their motivations and lifestyles would differ accordingly. So if lifestyles and even identities – which, even if granted a temporal dimension, stem from the sense and perception of one's position in social space – are not a simple matter of choice by increasingly autonomous agents at all but based on the classed habitus, does this mean they cannot be 'reflexive' in the way Giddens describes and that, therefore his theory should be dismissed outright? Not necessarily, for his overall position can, in fact, be reformulated along the following lines: owing to what he describes as the decline of traditional modes of practice and globalization, the lifeworlds of individuals have become suffused with new experiences and information on different ways of life and thus appear to offer new choices on how to live. Coupled with the pressures of consumerist individualism promulgated by the culture industries and political rhetoric in Britain over the last few decades, the propensity to 'reflexively' change and experiment with one's lifestyle choices *could then become incorporated in the habitus as an unreflexive disposition*, combining different levels of knowledge and attitudes built out of these new lifeworld experiences, once again in a way similar to that suggested by theorists of the 'reflexive habitus' (Sweetman, 2003).

However, it is not hard to imagine that new experiences and information flows would be distributed remarkably unevenly into lifeworlds on the basis of, firstly, the latter's material and cultural structuring (for example, access to the Internet, association with cosmopolitan significant others) and, derivatively, the practices and pursuits issuing from the agent's habitus: which television channels and programmes they watch, what they look up on the Internet, which newspapers and magazines they read and so on. Even if they do permeate the lifeworld they would be interpreted according to the extant perceptual schemes of the habitus. Furthermore, it seems likely that not only would a certain amount of both economic and cultural capital be required to realize a fully 'reflexive' pursuit of different lifestyles but also, because of this fact, the reflexive construction of one's lifestyle would for the most part be perceived through the lens of the habitus of those with less capital as 'not for the likes of us'.[13] To put it in a nutshell, the reflexive habitus might well be a preserve largely of those more distant from necessity, with the dominated remaining rather more 'univorous' in their consumption patterns, even if, recalling the principle of symbolic violence, they recognize that reflexivity is a legitimate or desirable style of life.

Conclusion: Themes for examination

Reassessing the theories of reflexivity from a phenomeno-Bourdieusian standpoint does not lead to the inevitable conclusion that all should be rejected outright. Instead we are left with a multitude of themes and possibilities that could still signal social changes broadly resembling those outlined by Beck and the others, albeit re-articulated in a different conceptual vocabulary in which the concept of class is both more subtle and more stubborn, and which must ultimately be confirmed or refuted through empirical research. It is worth, in this concluding section, setting up the following analysis by concisely recapitulating some of these themes and prospective scenarios and their order of investigation.

Only once the details of the interviewees' objective positions and trajectories have been dissected will we be able to comprehend their schemes of perception, so the empirical analysis begins with the examination of journeys through the social structure and, because the past undergirds all subsequent experience and action, childhood and education. Clearly the basic task in Chapter 4, the first of the empirical chapters, then, is to establish whether Bourdieu's theory of educational reproduction, with its focus on the transmission of capital and the unacknowledged circumscription of choice, still has any purchase, however modified by educational reform, or whether, conversely, education and the new opportunities available demand some form of reflexivity, especially when transitions are made to either post-compulsory education or employment. Chapter 5 will then pick up the interviewees' life courses from their entry points into the labour market to their biographical situations at the time of the research, with the intention of discerning whether career moves are as erratic as Beck and the others claim and motored by a disposition for reflexivity generated by current social conditions, or more stable and unreflexively guided by the objectified and internalized constraints of capital. In both cases it will be crucial to ascertain whether new processes and experiences are stratified in some way by position in social space.

With the structural dimension of class explored, the analysis will move on to the symbolic dimension, the constitutive practices of which can be partitioned into several layers: the unarticulated symbolic differentiation of lifestyle patterns, or *class practices*, through the often fuzzy and confused sense of difference and similarity, or *social identity*, to the explicit discourse of 'class' and its place within typification schemes and political propensities, or *class talk* and *class politics*. Chapter 6, the first of the chapters devoted to the symbolic realm, will cover lifestyles and identities and attempt to determine whether lifestyle practices remain explicable by trajectory and position and whether agents still display some form of 'class sense' based on the typification and reading of signs and behaviour as they narrate their biographies and reconstruct their lifeworlds, or whether this has dissolved in a sea of individualized and reflexive consumption practices. After that, Chapter 7 must establish whether people talk about 'class' and conceive and label others and their practices in these terms, whether and how this categorization has any salience and coherent meaning for them, and whether they use it as a framework for interpreting social injustices and political projects or whether individualized concerns, as Bauman suggests, or reflexively considered lifestyle issues, as Giddens claims, have largely taken its place. Only after all this will we be able to pronounce with some degree of certainty whether class lives on and is fit and well or, conversely, whether it should be interred once and for all.

Part II
Searching for the Reflexive Worker

4
Educational Reproduction Today

If freedom is nothing but 'the power of living as we choose', reasoned Epictetus, then 'the educated only are free'. Two millennia later, the reflexivity theorists would add that because more and more citizens of the affluent West are being educated to ever higher levels, then that freedom – that compulsory power to choose how to live – is diffusing throughout the population and battering down the barriers of old. A contentious claim for sure, but it has to be acknowledged that the restructuring of the education system the proponents of reflexivity take as their point of departure is well documented. As Western societies have increasingly de-industrialized and turned to knowledge and service provision as the principal activity of their economies, coupled with the birth of a globally hegemonic neo-liberal discourse that emphasizes the cultivation of human capital, post-compulsory and tertiary education systems across the world have indeed begun to mushroom and engulf larger tracts of their populations (OECD, 2008). The supposed corollary is that further and higher education, and the new experiences and options these deliver into lifeworlds, are no longer the exclusive preserve of an elite few. Instead, across the nations of Europe, North America and the Antipodes particularly, they are no less than *mass* phenomena (Halsey, 2000; Trow, 2005). In the UK this has been evidenced in the successive waves of expansion and increased student numbers ever since the Robbins Report in 1963, with the transformation of polytechnics into universities in 1992 and the drive for 'widening participation' by the New Labour government being significant recent developments (Archer et al., 2003).

Yet it is not just the spread of post-compulsory education alone that is thought to have induced reflexivity, but the changing *content* and *delivery* of pedagogic practice as well. There was, in the 1980s and the

1990s, a significant swing towards the right of the political spectrum in most Western nations and with it a growing emphasis on individualism and unbridled market forces, with the latter not only dictating the role and substance of education but serving as its model. Thus we have witnessed the marketization of state education to allow greater parental choice over school selection and competition, the logic being that standards will increase in the bid for student numbers in the same way as the private school system, which continues to churn out academic successes; the diversification of educational delivery, including 'lifelong learning' schemes, and the emergence of careers advice services like Connexions in the UK, both of which are conceived not only to facilitate 'free choice' in an ever diversifying labour market but to equip and re-equip individuals with the skills required by the capricious economy; and the appearance of a trope of 'personalized learning' in which learning plans are tailored to the needs, 'abilities' and interests of the individual child (see Brown and Lauder, 1996; Ball, 2008). Surely, Beck and the others would say, whatever their ideological provenance, these political interventions have served to induce choice and deliberation across the board, with even the option of manual labour having to be openly negotiated and justified *vis-à-vis* alternatives, while the perception of oneself as an autonomous, self-determining creature is encouraged from the start?

But a closer look at the statistics gives cause for caution. The prevailing finding would seem to be that while attendance and performance in post-compulsory education and higher education has moderately increased in absolute terms among the most disadvantaged in the last thirty years or so to match the expansion of education, the *relative* rates of participation, that is, the differential odds of attendance between positions in social space, have remained largely unchanged. This is a remarkably robust finding, unearthed with an impressive variety of indices of disadvantage and inequality (see Archer et al., 2003; HEFCE, 2005; Furlong and Cartmel, 2007: 27–30), but also a fairly widespread one transcending national specificities (OECD, 2008: 136ff). This means that the UK's stark figures – 44 per cent of eighteen-year-olds with professional parents continuing on to higher education compared with just 13 per cent of those with parents in routine occupations (ONS, 2006: 40) – are therefore rather representative.

So the broad patterns of inequality are still clearly in effect – they are, as Goldthorpe, following Merton, likes to say, an 'established phenomenon'. Yet if we are to fairly and satisfactorily adjudicate the theories in question we need to determine what the statistics, without unwarranted speculation, cannot: whether it is *reflexivity that produces these patterns*, whether people have to explicitly deliberate and assess a full range of options and select the course best suited to an individually conceived pathway to self-realization whatever that choice is, or whether class as developed in the previous chapter, with its emphasis on habitus, the taken-for-granted and mundane consciousness, shapes outcomes and choices. To do this effectively, educational trajectories must be unravelled and explored and the effects of parental capital transmission on school performance, consequent orientation to the school and then post-sixteen decision-making duly noted among the interviewees from dominant and dominated origins in social space. If the educational experience overruns differences of capital and generates orientations and deliberations that are uncoupled from class processes, then the reflexivity theorists can consider themselves vindicated. If, on the other hand, the iniquitous distribution of the capitals required to succeed – cultural, but also economic – invariably differentiates performance, experience and the valuation of schooling despite the changes in the social context, with the capital they depart with as a result then shaping future life courses by delimiting access to a particular field of possibles, then the continued potency of class will be demonstrated. In fact (at the expense of those who like their sociology to read like suspense thrillers), the following analysis will set out to demonstrate that the latter reality prevails, though not without important shifts in its manner, among not only the dominant and dominated but even the trickle of social space travellers who would otherwise be hailed as evidence of the retreat of class.

Symbolic mastery and the love of learning

For those thrown into lifeworlds where all the advantages in the world are simply 'there', in Heidegger's sense, the experience of and performance within the school are uniformly patterned and the subjective field of possibles tightly circumscribed around an entirely stable and taken-for-granted – that is, *unreflexive* – path to privilege. The causal chain in this Bourdieusian sequence begins with the dominant interviewees' parents, all affluent and equipped with the institutionalized markers of cultural capital (undergraduate and postgraduate degrees, A levels from a time when they were rarer and such like) to varying degrees, and their orientation to the educational system and process. No doubt based on an inculcated desire for their child to thrive (or 'reproduce' their social position) given the doxic construction of 'the good life' in capitalist societies and its particular spin by class fraction (accumulating wealth or knowledge), an awareness of what is necessary for this given their own experience – the 'information on the educational system' mentioned by Bourdieu and Passeron (1990: end diagram, n5) and examined by Devine (2004) – and a tacit protention (upon which projects are built) that this is probable, or at least realizable, given the stocks of capital they can invest (cf. Allatt, 1993; Reay, 1998a; Lareau, 2000, 2003; Vincent and Ball, 2007), each and every one of them was described as displaying a tangible and committed *involvement and interest in* and *encouragement of* school success – which, as we will see later, is a far from universal phenomenon. Typical descriptions rendered them as 'really encouraging and interested' (Karen, 28, junior doctor and daughter of a teacher and a social worker),[1] 'very pro-education' (Abby, 27, languages teacher, daughter of an affluent IT businessman and a teacher), 'wanting the best possible education' for their children (Isabelle, 26, NHS scientist and daughter of an NHS scientist and a teacher), wanting them 'to achieve and have every possible opportunity' (Courtney, 32, managing director of an independent film festival and daughter to a civil engineer and a computer analyst), demanding their children be more 'stretched' at school (Abby), always 'nagging' them to do their schoolwork (Rebecca, 30, human resources adviser and daughter to a cathedral dean and a teacher), rewarding success with toys and confectionary (Nigel, 45, university Reader, son of a doctor), buying extra books for their children's schooling (Emily, 45, personal assistant, daughter of affluent publicans) and attending courses solely for the purpose of augmenting their capacity to assist (Barry, 47, project manager and lecturer, son of an accounts worker and a school secretary).

These described attitudes, exhortations and behaviours, as signifiers of the orientations to the world framing their early lifeworlds, inevitably sedimented into the interviewees' own perceptions, behaviour and orientations within the schooling system, furnishing a doxic intuition of what is 'normal' or 'expected' or of 'what is done' in tune with the pedagogic ethos of the schooling system (disciplined study and such like). They were, therefore, 'good kids' who always did their homework (Isabelle), 'worked hard' (Karen), 'got on with it' (Sean, 36, software developer and son of an architect and an artist), were 'diligent' (Adrian, 40, wealthy solicitor and partner, son of a dentist and a teacher) and refrained from 'slacking off or messing around' (Nigel). This disposition towards schooling – this feeling 'at home' in the school environment (Bourdieu and Passeron, 1979: 13) – is not, of course, *necessarily* exclusive to the dominant but could be held by dominated individuals as a product of some complex of differentiating experience such as education-conscious parents or a drive to self-betterment (cf. Devine, 2004; Walkerdine et al., 1999), yet it yields less fruit if parents do not at the same time, like those studied here, *transmit the ample cultural capital* – as early ability to master the academic abstraction of symbols and principles – to *fuel* it. This was achieved via two modes of diffusion already hypothesized by Sullivan (2007) in her bid to carve cultural capital into multiple categories for the purposes of statistical operationalization. Firstly there is the more obvious, *active* form of transmission, that is, concentrated and capable assistance with the tasks set by their children's teachers throughout their education – something that nearly all of the advantaged interviewees reported:

[...] my mum did loads of my homework for me, which was very helpful! She joked that she did all of our A levels and all of our degrees!

(Rachael, 26, daughter of farm-owning social care workers)

I seem to recall my mother helping us out quite a lot when I was at primary school in terms of just helping us [and] just asking questions and getting me over that maths hurdle I thought I suffered from.

(Adrian)

[my mum would] sit me down and we'd do homework and she knew what I was supposed to be doing and what I had to be getting on with, so yeah, she would help me.

(Abby)

Mothers loomed large in this phenomenon, as Reay (1998a) has already explored, but fathers also contributed to this effort which, in training the mind to recognize and apply connections, associations and formulations, implants to varying degrees (depending on parental ability and their relationship with the child) the forms of symbolic knowledge demanded by the school and garnering the grades that act as marks of distinction.

Despite being directly related to school performance, however, active capital endowment is but one plank in the laying of foundations for academic success. A second, perhaps more fundamental, *passive* modality of cultural capital transmission exists in which the surrounding experiential fabric of the everyday lifeworld as shaped by the practices of parents with plentiful capital stocks, quite apart from any direct involvement in activities emanating from the school itself, sediments into the child and nurtures capital and capital-generating dispositions. The following are demonstrative:

> Yeah, she [my grandmother] must've been [one of the first women to go to university], and certainly to sort of read a, you know read a science. I mean it used to make us laugh cos all her, in her cupboards at home she wouldn't have the names of things she'd have the chemical formulae on them.
>
> (Elizabeth, 39, computer programmer, daughter
of a cattle and beef farmer and a teacher)

> But I think it helped that while I was at school my father was also studying for medical, further medical qualifications. So you know, we'd all sit in the same room and do our homework together, which was nice, you know, it felt it was natural to do this, sit there for an hour or an hour and a half or whatever it was in silence.
>
> (Nigel)

> She'd [my mother] spend a lot of time testing me on spellings and talking about – I remember all the time, it'd be like driving home and work out whether something was a noun, adjective or verb.
>
> (Abby)

> We'd always discuss, every evening meal, we'd talk about what we were doing, what I was studying. So there was always a very lively discussion every evening about what we were doing and you'd learn quite a lot through that. [We'd discuss] the topical issues of the day. For example, if rubbish was piled up in the street, why was that happening, what should be done to resolve it. [...] We would discuss topical things in the news or the newspapers at the time. If an election was coming up we would discuss that and what the political parties stood for. [...] I certainly knew what I thought about things and the reason why I held that view.
>
> (Nancy, 46, barrister, daughter of a director of a large
company and a teacher)

These snippets, not to mention the frequently reported visits by all to art galleries, museums, stately homes and theatre, discussion of art and photography, immersion in a world of books rather than television and the plethora of activities undertaken – lessons in singing, dancing, ballet, horse riding, playing musical instruments, gymnastics, tennis and so on – all with or at the behest of parents or other close kin, demonstrate the two principal forms of passive transmission (cf. Lareau, 2000, 2003): ongoing parent–child interactions pre-reflexively – that is, without conscious connection to educational advancement but merely as 'what one does' – centred on everyday learning and the use of reason; and, secondly, the recurrent routines, practices and experiences constituting time–space paths that include specific valued cultural activities but also more basically the taken-for-granted ways of acting and thinking that sink into the child's practice.[2] So, in the above cases, Elizabeth's existence in a lifeworld where (*inter alia*) chemical formulae, and an *interest* in chemical formulae, are part and parcel of quotidian life is bound to have left its residue in her stock of knowledge and, specifically, to have deposited socially valued contents within it; the everyday learning practices foisted upon Abby by her mother, as a use of capital to produce capital, inevitably heightened her ability; the routines established by Nigel's father in his effort to augment his own capital induced disciplined study as a part of his offspring's natural attitude; and, finally, Nancy's family habitually demanded and therefore fostered the use of reason, debate, questioning and the connection of concrete experience with abstract issues, ensuring both a capacity and confidence in articulating views on a range of topics.

But this is not a tale of cultural capital alone, for economic capital and, through what can be safely (given the theoretical conditions laid out in Chapter 3) described as parental *strategies*, its investment and conversion also conferred substantial advantages. Principally this was through the purchase of private education, something a surprisingly large proportion of the dominant interviewees either experienced or nearly experienced as their parents considered their options. This was not construed as an effortless commitment on their parents' behalf – many of the interviewees described with lament how they had to forgo holidays, new clothes and the other goods and practices common among their contemporaries at school – but they still possessed the ample capital, often aided by scholarships won through the transmission of cultural capital, to pay school fees and live comfortably. In any case, the effects of private schooling – its ethos and pedagogic practices – in shaping the orientations and performance of the interviewees cannot be underestimated: as Debra Roker (1993) has so clearly demonstrated, the assumption and expectation of hard work and educational success are communicated to pupils from day one, written into the practices of the school and sustained by the dispositions ('motivation', for example) and induced self-belief of the pupils themselves. In the words of the interviewees, verbalizing what was at the time the non-verbalized backcloth to life, in these institutions there are such 'great expectations' (Abby) – usually of boarding school or, in older times, grammar school among private primary schools and university among private secondary schools – and assumptions and practices oriented around them (for example, the oft-described strict yet engaged and concerned teaching) that 'you don't have any choice about working [...] the ethos of the school was that "of course you did your homework", there was, is wasn't an option, I suppose' (Elizabeth).

Consequently, not only were there 'no big discipline issues' or disruptions (Abby), but the library and schoolwork were the default options in free periods (Nancy), pupils motivated each other to succeed, were 'pushed and stretched' and were astounded and distressed when anything less than the highest marks were achieved (Abby), making them all the better receptacles in which to pour the 'academic' modes of knowledge, as many of the interviews described them, they specialized in, including 'French verbs at the age of seven' (Debbie, 36, writer and broadcaster, daughter of a wealthy entrepreneur and a housewife), Latin, piano, the arts and so on.[3]

It is no surprise, therefore, that, in combination with the ingrained diligence from parental expectations and assimilated capital, all the dominant interviewees developed not only an ability to manipulate and play with forms of knowledge, or a *symbolic mastery*, but, with that, something like a *love of learning*, that is, the translation of academic pursuits into avenues of self-realization. If the first of these is witnessed in their generally high grades (multiple top marks in O levels or GCSEs) and self-descriptions as 'doing well', 'excelling' or being 'successful' as well, the second is clearly evident in the ubiquitous description of school work as something that they 'enjoyed' or 'liked', found 'interesting', 'stimulating', 'fascinating' or 'fun', whether that be languages (Claire), law (Nancy), media studies (Rachael), English (Rebecca), music (the 'passion' of Jackie, 38, project manager, daughter of a manager and a teacher), physics (Barry) or some other cerebral passage to pleasure.

Higher education as 'the natural thing'

Having assimilated the cultural capital inhering in their surrounds, developed the dispositions endorsed and rewarded by the education system, notched up top marks in their exams and acquired the formal requirements to enrol in post-compulsory academic study, the stage was set for the crucial decision-making moment bearing on future trajectories and, one might expect, the feted appearance of reflexivity. Yet, in sheer numbers, nearly every dominant interviewee lived up to their statistical odds and opted for further study and, if we unpick the rationalizations of this transition and responses in an attempt to probe the range of options considered, the causal mechanisms underpinning this become clear quickly enough.[4] When queried whether any option other than the academic route, A levels, had been considered, the incredulous interviewees, whether old or young, male or female, promptly answered in the negative, affirming that they were 'always gonna do A levels' (Karen), that 'it went without saying' (Emily), that the prospect of leaving school 'never really occurred' to them (Barry), was 'never considered' (Jackie) and never broached in the familial environment or, especially for those at private school, among peers, teachers or careers advisers. Isabelle gives a fuller picture in her description of her transition from a private all-girls school to a mixed-sex sixth form at a different institution and captures in a sentence the underlying reality in all cases:

WA: So did you consider not doing A levels?

Isabelle: Not really I suppose, no. But I suppose it was just that pretty much everyone in my school was gonna go and do A levels and I also had the prospect of going to this school that I knew I was going to have a really good time. So *it just seemed like the natural thing, I don't think it really ever crossed my mind*, although friends that I had out of school weren't all going to do A levels, but they were all going into some sort of further training or study or something so, yeah don't think it was ever really an issue. (emphasis added)

In other words, the continuation of education (or 'training') was 'the natural thing' to do given the resources inhering in one's position and the expectations and exhortations filtering into a lifeworld rich with cultural capital. In stark contrast to the thesis being tested *there was little reflexivity here at all*. No other options ever 'crossed her mind' – they were, in other words, excluded *a priori* from consciousness as unthinkable given the field of possible trajectories open and the dispositions towards academic study instilled. This is brought home all the more by Jackie's shock and disbelief, mirrored in the accounts of others, at a contemporary who rebuffed schooling and left at sixteen – the exception, so to speak, that proves the rule or, more accurately, the habitus:

Jackie: No, never considered it. Very few people didn't [do A levels]. People went to different places, some people went to sixth form college, or to technical college. One or two girls left, one girl left without doing any O levels, which everybody was just really shocked by. She left at sixteen, she said 'I'm not coming anymore, I don't have to go to school.' I couldn't believe her parents let her, I just couldn't believe it!

WA: What do you think your parents would have done?

Jackie: I think there'd have been a bit of a scene actually. I think there'd have been a horrible, horrible scene. It's difficult, isn't it? I mean you can't force somebody of sixteen to go to school. I don't know what they'd have done. It was never – it just wouldn't have happened. [...] the majority of people stayed, and I think with a grammar school the majority of people stay. It's kind of expected you will, so you're not given a raft of options about what to do, it's kind of just assumed almost, that you'll stay.

Vitally the self-same process saw nearly every dominant interviewee proceed straight on after A levels (or their equivalent in different national contexts) through the increasingly central sanctifier of social division: higher education (cf. Reay et al., 2005). The same language abounds, with respondents reporting that there was 'no other option' (Humayun, 31, councillor, son of a Bangladeshi army officer and a housewife), that they 'didn't even think about it' (Helen, 42, business support manager, daughter of a sales director and a housewife), that it was something they simply 'assumed' they would do (Jackie) and that, in words analogous to Isabelle's on A levels, 'it was kind of natural progression to' go on to university (Mark, 35, computer programmer and son of a university professor and a teacher). The interviewees' own proffered explanations for this suite of assumptions hinge upon the prevalent expectations of the 'normal' among family, peers and the private school – 'everyone in my family had gone' or 'everyone at my school went' were frequent refrains – all of which, they were keen to point out in response to probing, were never *explicitly* stated but encrypted in the discursive and practical parameters of their lifeworlds:

[university] was always [my parents'] plan, so. It was never put to me as in 'you're going' [...] It was also sort of 'well when you're at university...' things like that so, it was kind of like that but no it wasn't 'you're going'.

(Claire)

It's funny, I don't ever remember having a discussion about that [whether I would go to university or not]. Any discussion of that type would probably have been more along 'what do you think you'll do at university' or 'I'm thinking of doing this at university, what do you think?', you know. So it sounds as if it was pretty hardwired in there.

(Nigel)

No, no one ever said to me 'you will go to university and that's that'. [...] I mean I think I got some good advice from school, it wasn't like 'academic and earn a lot of money', it was like 'you enjoy languages so maybe you'd like to do a degree in languages'. Wasn't kind of 'and that will give you this and this job', it was more based on your enjoyment in a subject and what was fulfilling to you.

(Abby)

From the very start these interviewees were surrounded by the expectation – furtively manifest in the sayings and doings of daily life and enabled only by the possession of capital – that higher education, the means of securing reproduction, was not just within the bounds of the probable, but was the *norm*, a 'commonplace destiny' and 'natural' future (Bourdieu and Passeron, 1979: 3–4), thus *unreflexively* fixing it within their own perceptions of the time to come and their conscious projects and leaving only where and what to study to be mulled within the subjective field of possibles, with the latter, as Abby hints, being invariably driven by the projects of self-realization witnessed earlier in precisely the same way as half a century before (Hyman, 1954). Cultural capital is key, of course, in setting the subjective perceptions of the possible and typical – being seen and seeing themselves as 'academic' (Karen) and 'up to it' (Nancy), that is, in possession of symbolic mastery, the environs and intellectual challenges of university-level education 'felt right' (Jackie) – but the undoubted role of other forms of capital must never be forgotten. There was, after all, enough *economic* capital in each case to allow the distance from exigency necessary for the suspension of paid employment and even, in many cases (where parents or other family members were especially affluent), enough to cover housing costs and living expenses and to pull university – and the range of privileged graduate-entry jobs that were subsequently taken up – even further into the realm of feasibility.

Practical mastery, rebellion and survival

Among the dominated the story is quite different. In opposition to the privileged inhabitants of social space it was the relative *lack* of capital absorbed from the early lifeworld and the corresponding habitus of both child and parent that profoundly shaped their experiences and assessments of schooling and provided the founding stones upon which their subsequent social stasis or short-range trajectories would be built. None of the dominated interviewees' parents, by virtue of their own educational trajectories, possessed much in the way of institutionalized cultural capital. Consequently they did not possess the means – nor, given their relative proximity to economic urgency and their own assessment, based on a subjective expectation of objective probability (cf. Bourdieu and Passeron, 1990: 155ff), of the utility of education, the time or inclination – to prepare their child for the demands of the education system and to complement them at home (cf. Evans, 2007). There was not the help and involvement reported by the capital-rich interviewees, not the persistent encouragement and high expectations, because their conditions of life dictated otherwise.

Thus, parental input was usually, and often bluntly, described as simply non-existent by both older and, despite recent moves among educational authorities to push parental involvement to the foreground (see Reay, 2005), younger interviewees fresh out of school – they 'didn't show a lot of interest in me in schooling' (Gary, 44, driving instructor, son of a poor, low-skill engineer and a secretary), had 'absolutely no involvement at all' (Yvonne, 42, driving instructor, daughter of a builder and a secretary), 'weren't fussed' about school performance (Tina, 18, apprentice painter, daughter of a bus driver and a cleaner) or, more universally, didn't 'push', 'support', 'encourage' or have 'aspirations for' them. At best there was a minimalist exhortation to attend school 'no matter what' (Joe, 35, technician, son of van driver and an administrator), complete homework alone (Gillian, 29, teaching assistant, daughter of a lorry driver and an office worker) and 'go as far as we could' (Jimmy, 47, postal worker, son of a window cleaner and a shop worker) given the little capital that could be passed on at home. Occasionally the underlying constraints of capital would be revealed, as in Jimmy's declaration that his parents 'didn't really help too much' with schoolwork because 'they used to work' so much to provide the necessities of life, Maureen's (43, housing adviser, daughter of a civil engineer and a home carer) assertion that her parents didn't have 'the time' to help with work or 'see it as really being important' because the demands of work, the pub and 'running about after us all' were too great, or Andy's (44, studio manager, son of an office worker and a shop worker) comment that his parents 'were so busy with the four other kids' – feeding, clothing, nursing – 'they had no idea when my exams were' and could devote no time to the discussion of his future. Perhaps the clearest example of the limits of parental economic and cultural capital, however, comes from Martin, a 34-year-old surgical nurse born to a heavy-plant driver and a housewife who left school at sixteen only to be upwardly mobile in the course of his subsequent work life:

> They would ask if I had homework to do and if I'd done it, and tell me to get on and do it, but not sit down with me and do it. Dad wasn't there, he was working, making the money, and mum was working as well. I'm not going to say I was left to my own devices, they would make sure I did the work that needed to be done, but I guess because of their own education they probably didn't want to sit down cos they probably thought I'd know more than they did anyway.

Without the transmission of cultural capital or the soaking up of privileged pro-educational orientations, the dominated interviewees were not equipped with the symbolic mastery necessary to succeed in school or the dispositions to conform to its demands and ethos. Nearly all lamented the school's focus on 'theory' (book learning, algebra, English, abstraction and so on) rather than 'practical' pursuits focused on the concrete or corporal dimension (fieldwork, technology, cookery, typing), and declaring oneself 'not very academic' or even 'thick' was common. Alienated by a system that esteems what they had little of, that 'gives training and information which can be fully received only by those who have had the training it does not give' (Bourdieu and Passeron, 1990: 128), many thus developed, in a way reminiscent of Willis' (1977) 'lads', oppositional attitudes based upon a vague valoriza- tion and pursuance of what they did possess: *practical knowledge* or *bodily prowess*. The school was soon written off as 'boring' (Gillian), a 'pointless waste of time' (Tina) that fails to prepare its recipients for 'jobs and the real world' (Zoe, 20, clerical worker, daughter of a postal worker and an office worker), and so the interviewees, searching for alternative sources of pleasure and worth, 'ran wild' (Gina, 30, learning support assistant, daughter of a bus driver and a postal worker), purposefully disrupted classes (Gillian), became 'tearaways' (Gary) and threw themselves into activities attributed little value by the educational system. For instance, Hannah, a 30-year-old administrator and daughter of a salesman and a charity worker, having been brought into a lifeworld characterized by a distinct scarcity of cultural capital, insufficient economic capital to invest and a supportive but *laissez-faire* parental disposition towards schooling, became, as she described it, 'disillusioned' with education. After portraying herself as not 'focused on school', 'not very academic' and 'a bit of a rebel' who was often disruptive yet as someone who has 'done very well in lots of different aspects' and with 'a passion for certain things, albeit a lot of it wasn't academic' (in her case horses, initiated by contact with pony-owning cousins), when asked what would have improved her experience of school she reflects:

> [...] schools back then, they were purely academic to be honest and there probably wasn't an awful lot they could've done. Whereas now, the schools that they're trying to create now have got more vocational opportunities as well, so that would have been more suited to me but at the time that wasn't the structure of the school. [...] *I just got the impression that they just wanted the ones, were really bothered or really interested in the kids that keep their heads down and do well, which I suppose is what you should do when you're at school but not everybody does.* (emphasis added)

In other words, Hannah resented the focused attention on those who 'kept their heads down and did well' – that is, those with the instilled disposition towards self-disciplined study in harmony with the pedagogic practice of a schooling system hinged on the inculcation and nurturance of cultural capital – and vaunted the vocational, the *practical*, that which involves a skilled mastery divorced from the symbolism and abstraction from which she had been excluded. This is, however, a valuation laced with internalized symbolic domination: on the one hand she is pleased that vocational training is being rolled out in schools, thereby unknowingly legitimating a sort of educational apartheid that seals the fate of children by channelling trajectories from a young age and perpetuating inequality, while on the other the dictates of symbolic violence – of seeing her own practice from the point of view of the dominant – means she immediately counters this with a hesitant recognition that self-disciplined study, the cultivation of capital, is what one '*should* do' at school, even if not everybody, including herself, manages to do this.

Similarly, having been questionably diagnosed with dyslexia after difficulties with reading and writing,[5] assigned to a raucous special needs class and labelled 'lazy and thick' and a 'failure' by teachers, Joe quickly turned away from and lost all interest in 'academic' matters at school and, like others of the sample in a similar position, instead turned to sport, in particular running (for others it was athletics, ice-skating or, probably most widespread beyond the sample, football), which, coupled with his ability to 'look after himself' physically, became his real source of 'self-motivation and pride' and 'respect within the school'; in short, school for him was 'not about learning, it [was] about *survival*' (emphasis added). The cultivation of the mind and engagement in academic abstraction, presupposing the possession of forms of knowledge he did not have as a product of his structured lifeworld, was thus supplanted with the cultivation of bodily aptitude and the investment of self-worth in the 'fighting strength' that Bourdieu (1984: 479) recognized as central to the system of values of (the male members of) the dominated class, the problem being that these do not accrue the resources translatable into symbolic capital, the key capable of opening the doors of society.

Work as 'the natural thing'

Few of the interviewees from the dominated section of social space completed post-compulsory education. Instead, and in line with the national statistical patterns, apprenticeships and full-time employment, after some false starts and initial periods of claiming unemployment benefits, were the usual destinations and, once again unfavourably to those heralding the spread of classless reflexivity, the journey there was undertaken along classed tracks. Sometimes, for example, there was an explicit perception of the capital constraints inscribed in their positions. Thus many of the interviewees, when queried whether they had considered staying on at school or college or even pursuing higher education, simply said they were 'not brainy enough' for A levels (Tina), that it was 'never an option' because they didn't have the 'top marks' that were perceived to be necessary (Trisha, 37, technician, daughter to an electrician and a cleaner) or that university was 'beyond them' as 'only clever people' went there (Gary). Even if they had possessed the cultural capital and the attendant grades to proceed on to post-compulsory education, furthermore, the restrictions of economic capital would have soon blocked their paths. Phil, a 33-year-old supervisor of manual workers, very much like a foreman, and son of an electrical engineer and a housewife, for example, though also declaring that he wasn't 'bright enough' to pursue higher education, states that he 'couldn't afford not to work' anyway, while Trisha claimed her parents simply 'couldn't afford' university and that

> the only reason I stayed on [at college] to get an ONC was through my work, and even then I had to pay for it by staying on very low wages for three years – or two years extra than what the normal was. And that was the only way I was able to pay for that.

Such recognized constraints, as an articulated sense of limits, are not, however, the prime mechanisms in the negotiation of school–work transitions, as if the interviewees had reflexively considered all options and acknowledged the restrictions of their situation, even for those who mention them. Instead, the movement into work was guided by the pre-reflexive expectations, orientations and valuations of the habitus, which is to say the constraints and limits of capital *internalized* as a product of recurrent experiences in the materially and culturally structured lifeworld. Such experiences emanate from two interweaving sources: on the one hand, the expectations among the consociates and institutions of the lifeworld that permeate everyday life, themselves grounded in the largely tacit assessment of what is reasonable given the levels and composition of capital etched into the situation, and on the other hand the anti-educational attitudes, described above, that were produced by the individual's educational performance and trajectory given their limits of capital (which Bourdieu would term *amor fati* – an induced 'love' of one's fate).

In its most basic guise, the expectations, interactions and experiences characteristic of the lifeworld as structured by its relational position in social space impact upon trajectories through the medium of parental advice or directives articulated to a greater or lesser degree (but certainly absorbed by the interviewees), and based on their own experience and assessment of the sensible, the achievable, the practical and the 'normal' given their positions and trajectories in social space, to 'get a trade' (Joe) or 'go into an apprenticeship' (Phil). Sometimes, however, this also shades over from parental expectations to the broader doxic features of the lifeworld:

Yvonne: [my parents' attitude was] I was to finish school and become a secretary cos *that's what girls did.*
WA: You got that impression?
Yvonne: Absolutely. That's what my sister was told to do and that's what I was […] (emphasis added)

Trisha: You either went to college or you went into work. Well, I was just expected to go into work.
WA: By your parents?
Trisha: By my parents and *by society really. It was just one of the natural things that everybody else did.* (emphasis added)

'That's what girls did', expected 'by society really' and – mirroring the scenario witnessed among the privileged – 'it was just one of the natural things that everybody else did': all these phrases reveal a perception, brought to the level of discourse only by probing, of the structures of normality furnished by the experiential milieu in which they lived their lives. Secretarial work was not what all girls did, but what all girls within Yvonne's parents' orbit of experience, that is, girls with a certain level of cultural and economic capital, did;[6] it was not 'society', in the sense of a national ethos, that expected Trisha, who was from an economically and culturally homogeneous rural area, to enter work, and it was not what everyone, as in the national population, did, but it was what 'society' *as she perceived it* within a particular materially articulated *Umwelt* expected and what people *within her experiential reach* – that is, people of a similar position – did.

In most cases, such lifeworld experiences operate to feed, supplement, complement and consolidate the negative valuation of the utility, difficulty and displeasure of education generated by the experiences of schooling. This orientation, ingrained into the habitus, is manifest in Joe's assessment of education as being not only uninteresting (particularly compared to his enthusiasm for running) but a 'waste of time' in his account of the fact that he had attempted to stay on at his school's sixth form purely in order to prolong the support he received for his running but

> realized within two weeks that it wasn't gonna work for me. I thought 'no, let's chip off and get a job' [...] I thought 'no, I'm just going to *waste two years here* putting energies into stuff *I'm not really interested in*, so let's get out there and do something'. (emphasis added)

Similarly, Tina, having been far from enamoured with school, voiced her assessment of continuing education:

> I don't wanna learn, I can't be bothered. I don't see the point in spending a long time doing something you're not going to do ever again, which is what ninety-nine per cent of people do. Yeah, it's a waste of time.

In a vivid display of facets of the dominated habitus, she articulates the negative perception of learning as a *chore* given the difficulties inherent in a position and familial life marked by minimal cultural capital and as *unpractical* ('a waste of time') given its distance from the realities of working life, or from *necessity* – there is no 'love of learning' as there was among those distant from the demands of economic necessity.

Interestingly, most of these classed trajectories occurred under either Conservative or New Labour governments and, hence, have been contextualized by an education system oriented by political discourses of choice, marketization, personal responsibility and 'negotiated individual learning plans' (see Gewirtz et al., 1995; Hatcher, 1998: 20; Ball et al., 2000; Ball, 2008) – exactly the kind of political context that Beck would argue individualizes school–work transitions. This does not mean that such policy measures have had no social consequences, however, or even that these consequences do not resemble processes identified by Beck and the others. Indeed, Tina's transition into an apprenticeship in painting and decorating, extensively recalled because it occurred so recently and remains fresh in her memory, appears to exhibit the hallmarks of institutionally aided reflexivity consonant with the theses of the reflexivity theorists:

Tina: I went through a stage where I liked drawing, so I thought 'well, what about painting?' I like drawing, and being creative, I thought I'd do that and then I can move on and perhaps be an interior designer or sommat like that. I ain't now, but that was the plan. That kinda went out the window.

WA: So was this something you thought much about?

Tina: I thought about it for about a month.

WA: Did you look at lots of information?

Tina: Yeah, mmm, mmm, mmm, mmm [indicating looking through all the options]. I thought 'yeah, this seems quite good'. Then I found out about the apprenticeship. How did I find out about the apprenticeship? I've no idea, but I found out about the apprenticeship. Oh I did it through school, through their Connexions person.

WA: So the school had a lot of information?

Tina: Yeah, they had a sort of, basically set it all up. Set me up with an interview over there, and all that lot.

There is, in this passage, evidence of Tina selecting her occupational trajectory on the basis of an *interest* (what would, in other words, be 'best for her' and aid some sense of self-realization) rather than any obvious exhortation, expectation or aspiration to follow family or communal traditions, of an emergence of a consciously conceived *plan* for the future (interior design), which has had to be revised, of a *diversity* of options and information supplied by organizations specifically instituted for the purpose of broadening (and demanding) individual 'choice' (Connexions) and of a concentrated and elaborated *deliberation* of these options until the final decision was taken.

It is, however, as edifying to consider the occupations Tina did examine in her deliberations as those she did *not* – she contemplated being a builder, a plasterer, a bricklayer and (her first love given first-hand experience) a butcher, working through each option and discounting it on the basis of implicit and explicit, embodied and symbolic constraints of gender (she was 'not strong enough' for the jobs or could not do them because she was 'a girl'), at all stages overtly rejecting what were constructed to be 'what everybody else did', that is, what *girls like her*, with stocks of capital like her, were perceived to do: 'office work' or 'shop work'. But because of the class-based processes already described she *never considered university* and she *never considered a profession*: they were *excluded from consciousness* and simply did not enter her deliberations or her (swiftly crushed) life plans – which are both merely a demonstration of mundane consciousness and projection within the subjective field of possibles, even if contextualized by a milieu of amplified information – as realistic possibilities. She did, in other words, seek to do something 'interesting', 'different' and 'creative', but *only within the limits carved out by the pre-reflexive assumptions of her capital-structured habitus*.

Other interviewees, both old and young, display similar realities: the *precise* occupation or trade taken up were not obviously steered by local or family tradition, as in hackneyed (but not untrue) images of sons following fathers down mines in the immediate post-war period (see Dennis et al., 1969), but worked out on a more contingent basis, with interviewees often saying they 'wondered' what to do (Jimmy), 'didn't know' (Zoe), balanced options (such as air hostess or hairdresser for Maureen) or searched for what they 'wanted' given the inclinations generated by idiosyncratic lifeworld experiences (Phil) – *yet pursuing work or an apprenticeship was still the 'natural thing' to do*. As Jimmy put it, 'I didn't even think about going to college or moving on. It was just a case of I was gonna leave and get a job straight away.' Furthermore, it is clear from the generic nature of the doxic expectations already seen above (get a trade, work, and so on) that, at the end of the day, the *type* of job mattered less than having *a* job among the dominated. This is as true for Tina – who sensed the expectation to 'get a job' and that it did not matter 'what I did' as long as it was 'something' and would readily turn her hand to 'something else' if needs be – as it was of Jimmy twenty years earlier:

> Jimmy: Then I seen this job in the job centre, took it up to the desk and they arranged the interview and I got the job through the interview. I was only sixteen then, just left school like.
> WA: Any reason why you chose that job to go for?
> Jimmy: No, it didn't jump out at me. It wasn't like 'this is going to be a great career', I just thought 'let's a get a job', and then let's build on life from there. It weren't nothing in there that was going to jump out and be a great career move.

All this is, in fact, remarkably reminiscent of Willis' (1977: 99–101) lads, who were themselves subjected to what were then new careers guidance schemes. Then as now occupational futures were presented as 'clouds of possibilities to be thought about and negotiated' (Beck and Beck-Gernsheim, 2002: 6) to greater and lesser degrees in mundane consciousness, though with a steadily altering and increasing assortment of options, pathways and information with more recent economic and political change – in that much Beck is right – but then as now the *sort* of information negotiated, the *spread* of possibilities thought about and the individual decisions enacted were grounded in and circumscribed by the pre-reflexive orientations furnished by capital, that is to say, by class habitus (cf. Ball et al., 2000; Lehmann, 2007).

Social space travellers: The upwardly mobile

Among the dominant and dominated alike, then, it appears that, inauspiciously for the reflexivity theorists, the relational effects of class and the social reproduction and stasis they bring persist within the social context of the new millennium. Yet such reproduction is not total, for among the interviewees there exists a small but not insignificant collection of individuals who, despite beginning life in the dominated sector of social space, ascended into more privileged positions through the course of their educational trajectories.[7] This is not, as cruder critics of Bourdieu-inspired analyses (such as Goldthorpe, 2007c) imply, somehow antithetical to the late thinker's perspective on class or, consequently, the phenomenologically fine-tuned version of it employed here. Indeed, as is amply demonstrated by his various passages on the changing locations of agents within the social space through individual investment and conversion of capital and the shifting preponderance and reward of particular occupational groups, and his factoring in of trajectory as the third dimension of social space to capture all of this, mobility is not only allowed for but part and parcel of the very definition of class. Of course the *extent* of mobility is always an empirical question, and statistically the prevailing finding, as already mentioned, indicates that only a small proportion of the dominated are being dug out of their social locations by the burgeoning educational system and propelled upwards. Nevertheless, a crucial curiosity is posed: what makes these social space travellers so different from the rest of those hailing from dominated positions? How do they manage to escape the common fate of their initial social neighbours and, importantly, do the kind of conditions described by Beck and the others play any part by, for example, prompting some form of constraint-surmounting reflexivity?

Initially, it would seem as if there is little if anything to separate the early conditions of life of the social space travellers and those who remain in the dominated sections, with residential milieus described using much the same vocabulary: if not red-brick railway terraces, postwar council houses or 1930s semis in 'depressed', 'poor' or 'grey' city areas pock-marked by lewd graffiti and populated by manual workers of varying standing, the unemployed or even drug addicts, then poor rural life. Their childhoods were perceived to be framed by a relative paucity of economic means – money was often described as 'tight', they had 'much less' money than their peers (Bernadette, 32, graphic designer, daughter of poor French farmers), were thus 'aware of money' as children and had to modify their behaviour (such as which shops they frequented) accordingly (Samuel, 35, surgeon, son of a prison officer and a housewife), and recreational goods enjoyed by other children, such as the celebrated Rubix cube, had to be foregone (Lisa, 34, HR officer, daughter of a draughtsman and a factory worker) – and none of their parents had attended university or offered any notable help with homework – with, for example, Tessa, a 28-year-old junior doctor and daughter to a lorry driver and a disabled mother, noticing that because her parents were 'not academic' they offered less assistance with schoolwork than those of members of her cohort with more capital. Yet all proceeded through A levels, with this path being considered not only possible or achievable but, in some cases, much like the dominant, as almost a *foregone conclusion*. As Tessa puts it, she was 'always going to do' A levels; 'there was no question about it' for Samuel; while Zack – a 28-year-old software engineer whose mother claimed disability benefit and whose step-father was an unsuccessful self-employed water-filler seller – claims it was 'expected' that he would continue education after GSCE level.

Not only that, but *university* was within their range of possibles too, presented to them as a realizable goal which, with advice from the school, in competition with other options and driven by what would be 'best for them' (in Zack's words, searching for what 'I wanted to do' and what 'took my fancy'), they either took for granted within the natural attitude or mulled in mundane consciousness. For Samuel, for example, there was again 'no question about it', and both he and Tessa considered a variety of options – such as teacher, scientist, occupational therapist – that assumed university education, even if the specific courses pursued were subject to balanced pragmatic considerations of job prospects, perceived aptitudes and what they found 'interesting' before the guidance of their schools intervened and guided them towards medicine. Similarly, Zack, though he had at least entertained the prospect of gaining work after school and feared it, reassured his anxious mother that 'yeah, course I'm going to uni', though this decision is rationalized in terms of his 'academic' A levels failing to 'point' to any practical application in a job and finding the prospect of employment 'unappealing'. And of course, as Beck and Archer would be quick to point out, all those who went to university were, like the dominant, 'disembedded' from their locale as a corollary, distanced socially and geographically from their early lifeworlds and subject to new experiences and conditionings, breaking the continuity with their classed pasts and sometimes leading to an experienced sense of disjuncture and unease with consociates from that period. As Zack put it:

> it's bizarre going back home and talking to people you knew and just seeing how much of a *divide's* in place even though it shouldn't be [...] it's like having done A levels distances you from a lot of the people you knew at school, because at that point a lot of people have diverted and gone for like vocational courses and kind of – well I say vocational, most of them were business studies which doesn't end up being too vocational I think, in practice. But they've gone more that way than A levels say, and you differentiate at that point, people have different things to talk about and start to have different attitudes I think. (emphasis added).

In all cases the protentive sense that university – the key breaking point in their trajectories – was within the realm of the possible, likely and desirable, and the attendant projects conceived and considered in consciousness, would appear to be rooted in two interacting factors. First of all, in line with the reflexivity theories, the expanded higher education sector and its consequences for school careers guidance and implicit and explicit familial expectations no doubt mean that university has become an option suggested to and considered by growing amounts of school pupils who would have struggled to compete for places in its more exclusive days. Secondly, there is the interviewees' self-perceived ability – they described themselves as 'smart' (Zack), 'hard working' (Tessa) or a combination of the two (Samuel), sometimes explained with reference to IQ and genetic inheritance, and thus saw it as almost self-evident that they would pursue academic studies. However, underlying the latter and contributing to the surrounding expectations and guidance is the real explanatory nub of the divergent trajectories: *parental strategies* – those multifaceted actions stemming from the complexes of dispositions manifest in a desire for offspring to attain symbolic recognition and (as part of that) material well-being which, in Britain and other Western societies, hinges on success within the educational system. These were engendered, it would seem, not as a universal feature of humanity, as Bourdieu sometimes implies, but by the slightly different positions, trajectories and attitudes of the parents in question compared to the those of the rest of the (immobile) dominated.

The strategies are evidenced by a number of intermeshed actions, behaviours and attitudes that set the parents apart, starting perhaps most fundamentally of all with the fact that they all, unlike the parents of the remainder of the dominated, seemed to place a high value upon educational success and encouraged it among their children. Lisa, whose mother had migrated from Ireland to Birmingham to work in a factory and whose draughtsman father was 'born in a council house', provides a typical example when she states that her parents saw education as 'really, really, really important' such that 'doing well at school was the only thing that mattered'. They may not have been able to assist with schoolwork and impart cultural capital themselves, given their own low holdings, but the message was clear: 'we were told that you had to do well at school or else!' Consequently, she reports that she 'worked hard' at school, and indeed attained entry to the nearby grammar school 'off the back' of her older siblings (the eleven-plus system had recently been abolished), and was thus subject to all the expectations and pedagogical modes of capital inculcation such institutions bring. Indeed, it was these expectations and the work ethic of the school, which 'reinforced [her] parents' values and vice-versa', that led her to do A levels and, eventually, follow all her older siblings on to university:

> [...] had I have not gone to the school that I went to I don't think I would've gone to university at all, and I think with the expectation

that I did go to university, if that wasn't there either I do honestly think I'd still be working in [supermarket chain where she worked as a teenager], or have about eight kids with God knows how many fathers. And I'm not saying that from a snobby point of view, but I know people who grew up on my road who've ended up in that sort of situation really.

Expanding on this last point, Lisa clearly recognized that her parents' attitudes and behaviours, which so evidently structured her everyday experience, were out of step with those of the parents of consociates in close geographical proximity:

> I think other parents were a lot more laid back really. [They said] 'oh yeah, no it's fine to go to the local comp'. There wasn't that pressure there at all to be different or to do well. I remember from, well ever since I can remember, like from being that big [indicates a short height with hands] that you had to do well at school, because the eldest in my family, the brother, is thirteen years older than me and he was at grammar school at the time. It was always there, always, always, always.

This account is paralleled by those provided by Tessa, Zack, Wendy (48, assistant head teacher, daughter of an avionics electrician and a market researcher) and Samuel – all their parents were perceivably encouraging of success at school even if unable to help once beyond a certain level, and all as a result reported working hard, to greater and lesser degrees, at school. But in the latter there is more. As part of their parents' strategies they were *purposefully sent to distant schools or colleges* with higher expectations and standards than local alternatives with 'bad reputations' (a common catchphrase), primarily because of the more affluent areas they served, sometimes accompanied by a residential relocation to 'better quality' or 'reasonably nice' housing, itself a testimony to the parents' strategies for themselves to progress (for analysis of the parent's-eye-view of school choice, see Gewirtz et al., 1995).

Precisely why the parents of the socially ascendant held such different views and undertook such different actions from the rest of the dominated, that is to say, pursued strategies of advancement, is not easy to ascertain given the low numbers of participants, but it is perhaps instructive to note that many of the parents actually occupied relatively high positions within the dominated section of the social space. Lisa's father, for example, was very much a skilled, less manual breed of worker as a draughtsman; Samuel's father was obviously senior enough within the prison service to afford to move to a 'middle class' area during his career; Wendy's father was a skilled avionics electrician affluent enough to own a 'nice' three-bed house in 'semi-rural' Kent and her mother had moved from hairdressing to market research within her working life; while Zack's mother, though unable to work owing to disability, was 'very bright indeed' and would have gone on to university had she not have been accused of cheating in an exam. Perhaps this accords well with the discourse of *self betterment* that accompanied many of the interviewees' understandings of their families' actions and wove through their life narratives – their parents populating that region of social space historically constructed as 'respectable working class', so close to the petit-bourgeois to the extent of shading into them and like them generally upwardly mobile into their positions, they display the disposition towards 'getting on', of perpetuating their past social movement into the future by continuing their familial trajectory through whatever means are appropriate, that forms a part of the greater internalization of bourgeois values long-said to have characterized this class fraction (cf. Bourdieu, 1984: 331ff; cf. Jackson and Marsden, 1962; Roberts, 2001: 83ff).[8] Thus Lisa, for example, in explaining her parents' motives, stated that 'it was the only way to get out of the life that you were in really, to be educated' and, a little later in the interview, that 'the only way to get on was through education really, and through hard work', while Samuel notes that his parents

> were always supportive and always really keen that we stayed at school cos they saw it as a means to a better end and the opportunities they hadn't had [...] They valued it and thought it was important cos they saw it as a way for us to better ourselves and that we had opportunities they didn't have.

Furthermore, in some cases there were clear, if at first hidden, advantages. To give just one example, Zack's grandmother, an office worker of forgotten rank with whom he had frequent contact, was not only depicted as 'quite smart' and 'clever with words' and 'always relatively concerned with how I was doing', but because of her disposition towards self-betterment through education (she 'put quite a lot of stock in academic achievement', and this was 'instilled in her by her own mother' as 'how we better ourselves'), two of her children – Zack's aunts – had attended university and become professionals (a teacher and an accountant) with considerable cultural capital. Having such people within his lifeworld who could not only inculcate certain forms of cultural capital but provide knowledge of and expectations towards higher education, as Bourdieu and Passeron (1979: 26) noted long ago, distinguished him from the other members of the dominated and no doubt increased his objective chances, and subjective perceptions, of progression through the education system. This did not go unnoticed by Zack himself, who acknowledged that 'they, I think, would probably have been inclined to hurry me towards A levels' and joked that 'they were probably there trying to teach me to read and stuff – "what's this word? Onomatopoeia – now spell that"'.

In sum, then, these individuals have seized upon the burgeoning higher education system and not only considered but pursued trajectories that, in a previous time, would have been closed to them, yet the answer to the question 'why them?' appears to remain, contrary to the ideas of Beck and the others, grounded in the relative advantages they possessed over their social neighbours and the strategies that issued therefrom. Having said this, however, the class origins of the upwardly mobile marked their entire trajectories and set them apart from the socially static dominant individuals they encountered through their educational careers and occupations as well. Because of their parents' inability to directly transmit cultural capital through focused or everyday learning, for example, their educational achievements were 'a conquest paid for in effort', as Bourdieu and Passeron (1979: 24) put it, that is, accomplished without the advantages, the ease, the lack of struggle and the assurance that come with parents holding cultural capital (with the partial exception of Zack, who saw himself as intellectually able but mischievous, for the reasons discussed above). Wendy, for instance, received encouragement from her parents but did not consider herself a 'leading light academically' at school and, while studying A levels, repeatedly failed and resat her exams, only finally, after a period of working and travelling, going on to and completing higher education (even if never able to 'break the sixty-five' per cent level in assessments) at a later date at a small institution devalued in the field of education. Likewise, Tessa describes how, because her parents were unable to help with schoolwork, she had to 'get on with it' by herself – displaying the 'defence' (or disposition) towards quiet determination with minimal 'fuss' noted among similarly placed girls by Walkerdine et al. (1999: 145–6), also evident in her resolve to temper the anguish her wayward brothers brought to her parents. Again, however, her whole trajectory, reliant on the self-accumulation of capital, was characterized by struggle. First of all, her hard-won A level grades were insufficient for direct entry into medicine, necessitating a year out of education – which she spent working and saving rather than travelling (an increasingly valued sign of 'broad horizons' and 'initiative' among employers) – before reapplication. Following this she had to undertake a pre-medical foundation course, setting her further behind those riding a seamless transition on the back of their inherited cultural capital, before finally, when at a red-brick university, struggling to keep up with the work, compounded by the economic necessity (itself exacerbated by the fact that, because she had to take a year out, she missed the final year of grants) compelling her to pursue part-time work throughout her degree course in factories, shops and the like, and eventually graduating without honours. All in all, it was, she said, 'hard work', and though she 'couldn't really say for sure' whether it affected her studies (despite admitting that 'maybe I could have done a bit better academically if I hadn't been working'), she was nonetheless jealous of those around her who did not have to work, and hence the bearer of deep-seated emotional impulses instilled by class difference, even if this was evidently mitigated by her disposition towards quiet determination and rationalization:

> Well, I guess [I was] jealous sometimes, cos who wouldn't be, I mean I would have preferred to be in a position where I didn't have to work and that I wasn't getting into debt, definitely. I think lots of my close friends actually didn't have to work at all and it was difficult sometimes. But that's just the way it was, and you know, *I got on*. On the plus side of it, I got a lot out of doing jobs and working, I know now that I picked up a lot more skills by working than people who go through university and they – certainly medicine – and they're quite isolated as to what it's like to actually work and work as a part of a team and all the rest of it. (emphasis added)

Further to this, Lisa captures another peculiarity of the upwardly mobile when asked about her parents' encouragement of particular career options:

> Lisa: I suppose it was just whatever you felt you were good at, or could do, but then had I have wanted to do something like drama or something, or art – I was never good at art, but just for example – then that wouldn't have been acceptable because that wouldn't get you a proper job. You know, had to be something that was, could get you a job, you know, that it was something quite sensible really, rather than something outlandish. And I think it was always that – I know it sounds contradictory – you've gotta do something that's good, but not that good.

So we wouldn't have been encouraged say, to do something like medicine, because that was out of your league, you know. It was in a sort of set area of acceptable sort of professions or routes that you would take. I know that sounds a bit weird, but yeah.

WA: Was that something that they kind of said or that you could sense?

Lisa: A bit of both basically, I think it was sort of implied but also through things that were said to my siblings. So for example my, when my brother did [...] my brother did the eleven-plus and did really, really well and was sort of earmarked to go to Cambridge, I mean he didn't in the end but it was like 'oh we can't do that, you can't go to Cambridge, that's far too, that's far too high up'. Do you know what I mean? It was sort of, you go to a red-brick or a polytechnic and do something really sort of middle of the road but you can't be exceptional basically. I mean maybe none of us were, but it was never the idea of that you can do whatever you want, whereas I know there's some sort of like liberal upbringings you're taught to sort of be whatever you want and do what you want to do and you're really great at everything kind of thing.

Expressed here, in a nutshell, are the limits of the conceivable range of possibles, the 'sensible', the 'set area of acceptable professions' held by Lisa's parents – illuminating in consciousness, like the beam from a torch, only a circumscribed arc of social space and leaving the rest in the unknown, unthinkable darkness – and conveyed to her explicitly and implicitly, based, in the end, on the limits of capital inhering in their structural location. The disposition towards self-betterment is underpinned by realism and, betraying their class roots, a commitment to the practical or vocational: on the one hand, the resources available and accrued can only go so far in reality (as we have seen) and, as such, set the bounds of perception, protention and projection, while on the other hand, the outcome must be a 'proper job', respectfully but not overly rewarded economically and symbolically, rather than, as is more the case with those distant from necessity, personal fulfilment. Samuel and Tessa displayed the same sense of realizable goals: remember both had considered a variety of mid-level, vocational courses (with Samuel explicitly vaunting the vocational) such as teaching and occupational therapy – it was only the schools they attended, at their parents' will, that intervened and set them on a different path in accordance with their institutionalized expectations.

Conclusion

Many of the shifts in the social architecture that Beck and the others perceive to be the underlying motors of individualization and increased reflexivity do, it seems, form part of the present context for individual educational trajectories. Post-compulsory education has expanded beyond its previous confines and, along with swallowing expanding tracts of the dominant for whom it has become part of their stable and taken-for-granted 'normal biographies' (Du Bois-Reymond, 1998; Reay et al., 2005: 33), offered an avenue of upward mobility to those who, in previous generations, would never have considered application, opening up the lifeworld to novel experiences of different physical spaces and diverse *modi vivendi* socially and geographically distant from the initial familial milieu. Alongside this 'disembedding mechanism', furthermore, there is even something to the idea that the transition from education to work among the dominated has become increasingly cast as a matter of individual choice in the face of multiple options, with schools and allied organizations supplying services and information and encouraging pupils to deliberate and decide what is 'best for them' in a post- (or de-) industrialized, de-traditionalized and neo-liberal climate in a way they simply would not have done in the immediate post-war years.

But this, as we have seen, does not portend the decline of class. Only if the sociologist failed to peer beneath the surface, to suspend the popular constructions and interpretations of social transformations and to operate with the relational conception of class could such a claim be made. For at all stages, at all times, and through all the above-identified phenomena, class processes continue to permeate the life courses of individuals, from their earliest lifeworld experiences right through their educational trajectories, not only where there is social reproduction but also where there is, against all odds, mobility through the social space. It sets the unreflexive bounds of conscious thought, the subjective field of possibles, in line with the range of moves objectively offered to them by the resources they hold and, thus, they read the discourse of choice and self-realization through the lens of a pre-reflexively anticipated generic path. The social and economic context may have changed, and the substantial processes through which inequalities born of capital distribution function may have been reconfigured, but the structure and effects of class remain as powerful and pervasive as ever.

5
Topographical Trajectories

So far the interviewees' life courses have disclosed an image somewhat at odds with Beck's (1992: 93) thesis that education involves simply 'choosing and planning one's own educational life course' and, while revealing social change, confirm that this hallowed institution, which all the theories under examination describe as a critical device in the genesis of reflexive agents, remains a pivotal instrument in the reproduction of class inequality. But what of life after education? Is it possible that, notwithstanding their classed paths through their school days and post-sixteen options, social conditions have intervened in the interviewees' subsequent trajectories in social space, prompted widespread reflexivity and reduced the effects of accumulated capital to nought? To answer this clamant question we need to set our sights on the relevant disembedding mechanism flagged time and again by the reflexivity theorists: the oft-heralded transformation of the world of work from a steady guarantor of lifelong employment and limited intra-generational mobility to a source of unmitigated flexibility, transience and erratic movement within the social topography of the class structure.

There are, in fact, two dimensions to this argument. One of them concerns the general patterning of employment histories and, in particular, posits the proliferation of what Brücker and Mayer (2005) have called 'de-standardization', where, even after the usual 'churning' of jobs and low-status starting points among young adults have been accounted for, people are increasingly spiralling in and out of temporary work and education or benefits through the life course and shuttling between disparate careers and, it might be inferred, positions in social space, thereby diminishing the impact of present position and classed past on the likely future. This claim is best investigated with statistical analysis and, indeed, several of those who have diligently done this have found at least some supporting data – most famously Leisering and Leibfried (1999) for Germany, but also Schroeder (2009) in the UK. Overall, however, the evidence is fairly firmly stacked against Beck and the others: there may have been a small increase in redundancies, temporary or 'non-normal' work situations (Beck, 2000a: 105) and career changes with the demands of neo-liberal policy – though in some instances claims to trans-*class* mobility may be artefacts of the crude categories used to measure it more than anything else – but the picture described by the reflexivity theorists is not only a tad exaggerated (Brückner and Mayer, 2005; Doogan, 2009) but, crucially, differentiated according to class anyway. Those with few economic means and little cultural or technical capital, for example, endure unstable and disjointed occupational histories to a far greater degree than their privileged peers (Fenton and Dermott, 2006; Goldthorpe and McKnight, 2006), and even where the capital-rich do switch jobs or careers for whatever reason they are unlikely to descend far in social space in the process (Schroeder, 2009).

Perhaps the theorists of reflexivity are blinded by the nostalgia often embedded in tales of breaks from the past (Strangleman, 2007) or have imbibed neo-liberal management-speak (Doogan, 2009). Either way, and even if Beck cannot make his mind up as to whether such findings mean much for his thesis – eagerly citing Leisering and Leibfried (Beck and Willms, 2004: 102) one moment, but then saying that the statistical patterns need not look any different the next – this is not enough on its own to refute the second dimension: reflexivity, and the notion that it, rather than classed dispositions, underlies whatever patterns there are by generating the choices made. It could still be that the proliferation of social travel, however limited and unevenly spread and whether enforced by imperatives of the economy or offered by new programmes for self-alteration, may have moulded a mindset more open to change, the consideration of multiple options, the desire or demand to break from past habituations, traditions and skills and the pursuit of some sense of self-realization in which the field of possibles structuring thought encompasses hitherto inconceivable far-flung corners of the social space among the majority of the population, even if the final outcomes remain stratified in some way. Only a deep excavation of occupationally mediated trajectories can assess whether this is the case and indeed, as the following will seek to demonstrate, reveal reflexivity to be a mirage reducible to a new manifestation of old class processes.

First impressions

The first task in this endeavour is to gauge the general level of 'de-standardization' among interviewees and, thus, the most likely nodes of reflexivity by observing the broad trends in their occupational histories and noting the incidence and character of job shifting. As it happens, many of the interviewees, no matter their position in social space, have indeed had their employment histories marred by recurrent insecurity, punctuated by redundancies of the ilk claimed to prompt reflexive self-awareness and scramble classed patterns of life chances, or disjointed by active job shifts in search of contentment, with some seizing on emerging opportunities to reskill through workplace or 'lifelong learning' schemes that presuppose and encourage a flexible and reflexive workforce (cf. Edwards et al., 2002). Many such shifts, furthermore, appear at first glance to be fairly drastic ruptures – for example, from teacher to business support manager (Helen), from shop worker to social worker (Sonia, 25, daughter of a social worker and a carer), from painter and decorator to software developer (Paul, 41, son of an electrician and a secretary), from leisure centre assistant to deputy headteacher (Wendy), and so on – and in some instances, fitting Archer's image of 'meta-reflexives', seem to forgo opportunities afforded in the occupational sphere by accumulated capital in favour of attempts to subordinate paid employment to or conflate it with lifestyle pursuits. In Claire's case, for example, there was a reliance on short periods of temporary work after university solely to fund travelling, and for Mark a readiness to take voluntary and low-pay conservation or holiday camp work despite his possession of a degree in geography in a bid to immerse himself in his interest in nature.

Even where such shifts have not occurred, moreover, there is a widespread *perception* among interviewees, even dominant ones in apparently secure occupations, that, as Bauman would expect, their current positions are far from set in stone. Abby, for example, having spent several years as a secondary level teacher and working up to department head, is open to the possibility of retraining into a new profession on the understanding that teaching is not, as it was for her mother a generation earlier, a 'job for life' given the constant perceived threat of redundancy, and indeed that career shifts are 'almost expected' nowadays as 'people are willing to take more risks', while Nancy, despite having been a well-paid barrister for many years, may retrain or move into property development in order to 'do something different and a bit more relaxing'. It is interesting to note, however, that Abby and Nancy have both had relatively stable professional careers, so it could be that they are, as Doogan (2009) suspects is the case among the populace, merely buying into the neo-liberal management rhetoric of occupational insecurity and flexibility when the facts, especially for professionals like them, do not measure up. In fact quite a few of the interviewees, especially among the dominant, have had rather more stable pathways than might be supposed if de-standardization were a widespread phenomenon, and, to counter the initial impression above, they actually *include* in their ranks Claire and Mark. After all, their unsettled movements at the outset of their working lives occupied, in reality, only a relatively brief period in their trajectories before being supplanted by commitments to conventional professional careers in business services and computer programming respectively.

Faux reflexivity

So, even if de-standardization is a limited affair among the population at large it seems as if we are dealing with a relatively mixed bunch here, providing us with an ideal platform for the main task in hand: examining the explanatory principles underlying occupational histories. Turning to this, the initial outlook would again, admittedly, appear favourable to the covey of class detractors: deliberation, planning, negotiation and active decisions, all in the quest for what is best 'for me', *seem common among dominant and dominated alike*. Within the *dominant* sector of social space, for example, there is Jackie, who, after her undergraduate and postgraduate degrees in music, worked for a music publisher and the Tate Gallery before then embarking on an MBA and shifting careers into personal assistance and 'knowledge management' in the private and public sector until that was cut short by redundancy, at which point she relocated to Bristol from London and became a project manager in the public sector. None of this automatically implies reflexivity, of course, but then she claims that she and her partner had 'thought about', 'plotted' and 'talked about' moving to Bristol from London 'for years', primarily – and here Giddens' thesis on the subordination of work to reflexive lifestyle choices receives support – on the basis of its perceived 'lifestyle' benefits, namely its reduced demand for commuting and hence greater opportunity for leisure time, as well as its classical and chamber music facilities (groups, orchestras and so on). It was, she continues, her redundancy, as Beck might expect, that prompted the move by presenting an 'opportunity'. Similarly, Karen's decisions to take up medicine rather than teaching after graduating in psychology and to become a GP in particular in order to balance her career with her desire for a family, a concern demonstrative of the new considerations necessitated by increased professionalization of women, were carefully mulled and planned – after all, she 'always thinks a lot about everything' – in an effort to pursue what she 'enjoys' and 'loves', namely 'looking after people'.

Both of these examples demonstrate the blending of self-realizing lifestyle-oriented choices with socially mediated biographical events – redundancy for Jackie, motherhood for Karen – that together characterize the 'fateful moments' described by Giddens, but a rather more complex picture that shades more into Bauman's pessimistic vision is provided by others. Take Debbie, for instance, a science writer and broadcaster in her thirties. On the one hand, she describes her leap from pharmaceuticals sales into science writing in a way that appears to convey the new sense of freedom said to be offered to citizens of liquid modernity, here concretized in the expanded differentiation and availability to mature students of postgraduate studies:

> after doing that for about three years I decided enough of sort of spinning the tale of the pharmo industry and just talking about one or two particular drugs, I wanted to broaden my – not knowledge, well knowledge and getting involved in a broader sense with medical writing, use of communication of science. So I went off and did a Masters [in science communication], and that's when the career changed.

But this newly occupied position is a double-edged sword: operating on a freelance basis – one of the employment modes hypothesized to be ever more present in place of secure and permanent work – the insecurities inscribed in the present demand short-term horizons and flexibility in the face of what is to come. When her perception of future probables, and whether they entail a perpetuation of the past, is probed, Debbie frankly replies:

> I just don't want to pin lots of hopes on something that might not happen. And the nature of my freelance work tends to just meander through life and through whatever crops up and the opportunities available.

She is far from alone. Barry, too, indicates that much of the apparent reflexivity that prevails is induced by social conditions when commenting on his redundancy from his previous work in knowledge management for a telecommunications company, his subsequent six-month period out of work and his taking up of consultancy work in the financial services sector: 'I could happily carry on with doing what I'm doing, but I guess something happens to make it change. I've not gone out trying to find new opportunities and change necessarily.' The difference, it would seem, is that Barry's previous long-term employment has not prepared him for the demand to change, whereas Debbie's work seems to stipulate that flexibility or reflexivity be a disposition, as Sweetman (2003) would argue.

Others have taken a third route, of the kind anticipated by Bauman: consciously (reflexively?) seeking to avoid the perceived insecurities of the labour market by training, or retraining, in avenues of work perceived as 'safe'. Sean and Sonia demonstrate this fraught bid to latch on to some trace of permanence in parallel fashion. Sonia insists that her choice to retrain in social work rather than psychology – after dissatisfaction in her low-level retail work had prompted 'umming and ahring' and got her 'thinking perhaps I could go back to college and do something else' – was based on the reasoning that

> I didn't want to do a psychology degree because I know people who have done that and not been able to get jobs. Ultimately, we need to have money to live, so I needed to do something that would get me a job at the end.

Sean, on the other hand, describes his transfer from photograph archivist in the media industry into software development and support, and his picking up of a Masters degree in the process, thus: 'I chose to do it cos I knew it was good transferable skills as in I could probably go anywhere and do it and hopefully get a relatively good job.' How successful such attempts to stave off insecurity are, though, is unclear. Sean feels his present job to be relatively secure, that 'there are other people that I think would be laid off before me', but that 'recently there's been a couple people being sacked and a few years ago there was another lull', and given the onset of the current recession

> it's one thing that would stop me for looking for a job now. If I went to a company now, places are definitely making cutbacks. In fact, if they're recruiting you hope it's because they have different work, but you never know, could be six months' time you disappear out the door.

Clearly, then, the strategy of trying to locate apparent safe havens is limited by the knowledge that they may not stay quite so secure for long. It is not only the dominant that relay active narratives and apparent episodes of reflexivity, though, but the *dominated* too. The majority, like the dominant, discuss occasions of active, deliberated choice and change regarding their present and future trajectories and, what is more, some of these appear to bear all the hallmarks of reflexivity in the sense of taking *oneself* as the focal point of life decisions. So Chris (33, customer services adviser, son of a health and safety advisor and a cleaner) tells of multiple moves through his occupational history in the idiom of the quest for self-realization (what 'I' want) and self-determination ('I' decided) at the heart of individualization (Beck and Beck-Gernsheim, 2002: 26): after being a nightshift worker at a large distribution centre, he realized 'I wasn't going anywhere, so *I decided I wanted a change'*. He thus moved on to courier work, but again:

> *I decided I didn't want to drive any more.* Sometimes it was just such a hard slog to get out of bed to go and do eight or nine hours worth or driving around everywhere in the country. If you don't feel like it, there's nothing worse than spending all your time driving. *So in the end I decided to move on.*

This was followed by call centre work for a car insurance company where '*I decided that I wanted to start my progression'* into supervisory roles, before being offered a middle management role at a furniture store, which, needless to say, he 'decided to take'. But then after leaving there and feeling frustrated at the lack of management opportunities in his subsequent work manning telephones again, this time with the British School of Motoring, he states: '*I decided to leave* and started my new job' at the call centre of a private health insurance company because he '*just decided I'd have a change and do something new'* (all emphases added).

Likewise, Eddie (37, caretaker, son of a mechanic and a housewife) says of his transition from being a leisure centre lifeguard – which he had been for sixteen years – that he was 'thinking of changing and doing something different' because he was no longer content with the 'hassle' he was receiving from children at the pool. But when he moved into low-skill work as an order picker for a frozen foods company, immediately 'I was looking for something else to do and I would've done anything to get out', 'I needed something else to do, I was getting bored with it and *I wasn't happy with it'* (emphasis added), so he left and, after one or two other jobs, eventually became a school caretaker.

In some cases, as with the dominant, the social root from which these apparently reflexive moments stem seem to have been the variety of 'personal trouble' such as divorce or redundancy that, according to Beck, has proliferated and spread across all sectors of society in reflexive modernity. For Joe, after his divorce and the subsequent collapse of his employment in the emotional fallout, he

> just thought 'right, okay, I'd better *search for what I want*, get on a path and hope to find something'. And *I made a decision I wanted* to get into building maintenance, and then started searching for roles like that. (emphasis added)

For others, it was redundancy that struck and, in Gary's and Dave's (51, lorry driver, son of a routine white-collar worker and a nurse) cases and seemingly favourable to Beck's and Bauman's thoughts on the new volatility of social hierarchies given the caprices of global capitalism, dislodged them from the managerial positions they had managed to acquire, and hence the routines and experiential hubs they had established, and prompted them to turn back on themselves in order to actively create new ones:

> So for the first time in my life I found myself actually without a job, and course I still had my [HGV] licence so I just did a bit of agency work for a while. You can always get agency work, just like that, pick up the phone and got offers coming in. And then I thought, 'well know what, I'll try a bit of foreign continental work' – always fancied that, you know, like going overseas. So did that for a couple of years, went all over Europe – Spain, France, Italy, even went down to Kosovo, supplied the British army down there.
>
> (Dave)

> In an ideal world I probably would [still] be a manager of an ice rink now and on good money, very good money, and bonus and all the rest of it. But because the path was cut off at that point [because of redundancy], then I was on the [metaphorical] road again, 'where do I go now? Where do I go now? Where do I go now? Come to the roundabout, I've got A, B, C or D, or exit one, two, three, four – which one do I take? Right, let's try one and let's go on and see what happens.'
>
> (Gary)

Others went in the opposite direction, describing their upwards travels in social space in terms of 'moments of clarity' after redundancy

(Martin) or a realization of how to attain 'self-actualization' (Wendy), but all these representatives of the dominated give the impression that, in Bauman's terminology, they have been forced to remould themselves and maintain flexibility in the face of fluid social structures and transient bonds, and it is, Beck would no doubt claim, the individual 'I', not what local or familial tradition or significant others dictate, that takes centre stage, assesses the options and enacts decisions in all cases.

Notice, however, that the language used so far has been deliberately guarded: all these findings *seem* or *appear* to demonstrate reflexivity in action. This is because, in reality, only the superficial level of personal narrative lends any credibility to the claims of the reflexivity theorists. When we begin to comb through the accounts in greater depth it soon becomes clear that underlying *each* and *every* event in the interviewees' occupational trajectories, whether dominant or dominated, are the continuing effects of structures of class difference in the form of two sources of social inertia: capital *objectified* and capital *embodied* (Bourdieu, 1984: 110). Together they reveal the active and considered decisions seen above not to be antecedentless sources of change but a *faux reflexivity*, that is, nothing more than mundane consciousness operating within the subjective field of possibles given class positions and dispositions but masquerading at the narrative level as action without limits or history.

Objectified capital

Capital objectified in *things*, whether money, goods, qualifications or consociates, does not come in neat categories, for each individual, possessing a unique configuration of capital, will have multiple concretizations of the different forms of resource available which allot advantages in some instances and impose impediments in others. Furthermore, these are, as will be observed later, usually adapted to, taken for granted and therefore the limits rather than the contents of consciousness when considering options in a given situation. Nevertheless, objectified capital does aggregate to form a totality affording the individual agent greater or lesser freedom and greater or fewer opportunities relative to others, and, occasionally, we can glimpse the more or less consciously grasped enablements providing the conditions of action for the dominant and, in the case of the dominated, the times when intervening experiences promote the projection of a specific course of action that bumps against the obdurate confines of their conditions of existence.

Beginning with the dominant, we can make out the multiple ways in which objectified capital underpins and contextualizes many supposedly reflexive decisions. Economic capital is perhaps the easiest to detect, its use presenting itself among the interviewees in three forms. Firstly, there are, as was anticipated when reformulating the reflexivity thesis in Chapter 3, numerous instances where decisions to undertake occupational change after redundancy were only made possible by substantial payouts and existing capital reserves. Take, for instance, Jackie and Barry, both of whom appeared to be exemplars of reflexivity above. The reality of the matter is that Jackie only had the freedom to consider and make the move to Bristol because

> I could just take the [redundancy] money and run, which I did, and civil service redundancy terms are *very generous* so I walked away with a *comfortable enough amount of money to go*. So I took the money and ran, and moved down here and looked [for work]. (emphasis added)

There was little pressure to gain employment forthwith to assuage material urgencies and demands, and in fact she had three comfortable months out of work looking for a job that matched her desires to remain in public services at a high level – in other words, to undertake minimal movement in social space. Likewise, Barry notes that his

> redundancy payment was quite good *to give a good buffer for basically a year*. In terms of the money, *you had a year to find another job before you're biting into savings.* (emphasis added)

Indeed, Barry had such liberty from necessity provided by his redundancy package and accumulated savings that he could opt to spend months doing 'bits of pieces I wanted to' like 'building work on our house' before he ventured back in to the workplace in a position broadly homologous with his previous occupation.

Secondly, there is economic capital by proxy – the monetary assistance provided by friends and family, especially parents, rich in economic capital that technically comes under the heading of social capital but which, because it manifests as pecuniary advantage, can be considered here. This tended to be drawn upon in two ways by dominant interviewees. On the one hand, it was notably common for them to tap parental capital stocks for a residential property purchase, either in the form of deposits or mortgage payments. This might not immediately seem to connect with work histories, other than the obvious point that home ownership provides a measure of wealth and security and thus boosts available options, but in fact it does in several ways. It allows, first of all, some respondents the freedom to pursue an advantageous occupational change dependent upon geographical mobility, as in the case of Emily, who relocated to Bristol with the assistance of her parents to maintain her upwards motion through the hierarchy in retail administration. It grants some the liberty, secondly, to meander between options in the search for self-realization, as for Nancy, whose father paid her mortgage for her when she opted to return to being a barrister, but 'at the bottom of the pile' and thus with a mismatch of income and outgoings, after leaving in dissatisfaction and spending time out as a lawyer for a public sector organization. In some instances, finally, 'property for use', to use John Scott's (1991: 65) terms, shades over into 'property for power', that is, a source of supplementary income heightening distance from necessity as the property is leased to others at profitable rates. This is true of a number of the privileged respondents, but its role as a foundation for projects of self-realization is nowhere clearer than in the case of Diane, whose father, a wealthy harbourmaster, has regularly covered the mortgage and bills on the large, expensive house she owns in the most affluent area of Bristol and which she lets out to fund her interest, currently being transformed into a business venture, in designing maps. In fact the residence her father's support has allowed her to retain grants such a level of economic freedom that the field of possibles opened to mundane consciousness is incredibly wide, clearly generating, or at least sustaining, a blasé disposition towards the present and future which, in the event of actualization in action, could easily be misperceived as reflexivity:

> And with my business now I can honestly say just see how it all goes, and give it my best shot, but if it doesn't work I expect I'll just turn around, especially if we're going into a recession now, rent the whole house out and go 'now I'm free, what should I do?' World's my oyster [...]

Diane's case points to another, more general way in which proxy economic capital generates dispositions and trajectories adjusted to distance from necessity: *the provision of safety nets*. Only because of his wife's comfortable income could Sean afford to take a year out of the labour market and retrain in the course mentioned above, for example, while the early attempts of Claire and Mark to subordinate or conflate their employment with leisure pursuits, far from being a denial of the opportunities furnished by capital in 'meta-reflexive' fashion, were in fact only made possible by the buffers (loans, paid bills, living costs) their affluent parents provided. Though Claire makes oblique reference to this by admitting that she lived with her parents while temping without any expectation of rent or contribution to the household – a luxury many poorer households would be unable to sustain – Mark sums this up most lucidly:

[...] my parents have never been short of money. Always say, they've always been there to help out with the university and, when in Bristol [doing low-pay jobs] *they've always been there as a fall-back.* (emphasis added)

Only their classed conditions of existence, therefore, allowed them to judge achievable, construct and follow projects of self-realization, rendering what appeared to be reflexivity little more than the disposition to pursue options within the field of possibles.

But fiscal resources are not alone in securing advantages for the dominant. The exploitation of *social capital*, more narrowly defined as contacts and networks providing reliable roads into new jobs, also maintains their social stasis within the social space through occupational change. Just to give three typical examples:

There's a lot of agencies around for contract work. I wasn't getting very far with them, and I knew about the organization I work for, they're a big financial services company in Bristol, and *eventually I got the role through a contact, another friend of mine who was working there who knew I was looking for work, got my CV in front of somebody and I got my first role.* So that's how it came about. I knew there were vacancies in various places but I wasn't getting put forward through the agencies, they didn't really seem to help. *It seemed to come much more from somebody knowing a vacancy somewhere and in a position to get your CV in front of someone.* So that's how that happened. [When that contract ended] I'd sent him an email saying I'm looking for another contract, and he was quite – he phoned me up and asked if I was interested in the role still. I said yeah, and then I got a call two hours later saying do you want to start a six-month contract next week?

(Barry, all emphases added)

And I was quite interested in information-knowledge management cos I'd always been dabbling with that, and a company were looking for a knowledge manager and didn't know where to go. For some reason they didn't go to the recruiters that deal with those kind of areas, they went to my friend at a personnel firm, who's a school friend, just completely out the blue said 'oh we need to recruit this post, you know we know we usually come to you for accountants, we don't know what to do about this', and she said *'I know exactly the person that can do that job.'* And I just went bowling along with my new-found confidence and just got the job. I was lucky. *If they'd gone to a, you know a proper agency, it probably would've gone to somebody else, more experienced. But that was kind of my lucky break.*

(Jackie, all emphases added)

Lisa: *Got a job through a friend working in London,* in their HR department. It was for a housing charity, so that was my first HR job, and was just sort of like their administrator.

WA: Which friend was that?

Lisa: A friend that I met at university, *a friend of a friend basically,* it was his wife. She was HR director, *so she sort of gave me a job essentially.* (all emphases added)

Had these respondents, and the numerous others like them, not known people with such power within their respective institutions – sometimes direct links, sometimes the weaker ties mediated by others famously explored by Granovetter (1973), but usually assembled when traversing institutions of privilege, especially educational ones – they would not have secured the work, and the positions in social space, they did when they did.[1]

All this has to be compared with the situation of the dominated, whose objectified capital stocks acted as constant roadblocks imprisoning them within their circumscribed segment of social space. In terms of cultural capital, for example, Andy was excluded from a desired occupation (housing officer) on the basis of his dearth of qualifications (institutionalized cultural capital), while Chris laments the fact that, when looking at management jobs in the local newspaper,

a lot of companies that want people will want people to a degree standard or will want people to a certain standard, and sometimes I look at it and think I could do that job, I've got the experience to do it, *but they want to see a bit of paper first of all.* [...] But sometimes you think, well, do I really need that bit of paper to prove I can do a job? I know that that bit of paper will tell them a lot about a person, but at the same time, sometimes you think that maybe you could do it as well, given the chance. Sometimes you wanna shake your fist and say 'I can do that job!' I could earn that kind of money, give me the chance to do it.

However, the usual manifestation of objectified capital is simply *the pressure of economic necessity* pressing certain options upon interviewees while pushing certain others beyond their reach. Examples abound, but just to name a few: Joe cannot afford a college course on air conditioning and refrigeration given the low economic capital inhering in his position, setting his projected goals ever more distant in the future; Hannah is resigned to a routine administrative job demanding less creativity and imagination than she would like because she needs to 'pay the bills'; Phil is currently unable to top up his HNC into a degree because he 'can't afford that'; Eddie had to give up work as a motorbike salesman on commission before long because he 'couldn't afford to stay there' given the low rates of return to the novice, and stresses that he has 'never [had] a long period without a job, cos I knew I had responsibilities' such as 'family, paying the mortgage, paying the bills, food et cetera' – compare this to Barry's luxurious break from work – and so on. It is, however, perhaps best captured by Doug, who, having had to leave his plastering apprenticeship prematurely and thus forgo the official qualification owing to 'pressure to earn money' exerted by his conditions of life, reflects on the reality of the situation he – and so many others in homologous positions – faces:

> [...] it's all a case of having to push yourself to make money, you know, you don't ever get the chance to sit back and think 'well if I do this', you know. It's a case of you've got to go and earn money [...] So your decisions ain't always down to yourself, it's kind of like you're looking at what you've gotta do, what you're pressured into doing.

In other words, the demands of material exigency, representative of a *class* of conditions of existence, effectively foreclosed not only the possibility of considering the kinds of projects of self-realization open to the dominant but even the pursuit of modest capital gains.

Embodied capital

Yet the consciously experienced constraints and enablements of objectified capital are not even the primary means through which conditions of existence direct work trajectories. That function falls to *embodied* capital – the capital internalized as a tacit *sense* of limits and opportunities guiding thought on future possibles by circumscribing the options that enter mundane consciousness and the weights attached to them. This operates with greater and lesser degrees of specificity. *Generic* embodied capital, for example, fosters a general perception of which directions and distances in social space are possible. This can issue from economic capital, as the pre-reflective sense of *affordability* widens or narrows the range of actions flowing through consciousness, evident most basically in the fact that people simply do not normally consider unrealistic avenues (as we saw with Tina in the last chapter), but the effects of generic embodied constraints and enablements are never more clearly revealed than when they derive from cultural capital, or perceived *ability*. Consider, for instance, the tendency among the dominant, generally absent among the dominated, to consider and pursue periods of retraining through university, including at postgraduate level. Abby, for example, when pushed on her willingness to leave teaching and reskill, states that she could 'perhaps' afford the substantial costs of doing another degree, but 'certainly another year' of training and, implied in her consideration of it as a realistic possibility and recalling the 'I-can-do-it-again' disposition described by Husserl and Schutz, perceives herself as capable of cultivating further cultural capital given her store of capacities and orientations already applied and amplified in previous rounds of higher education.

Compare this with the dominated, for whom the shortage of embodied cultural capital acts as an internalized barrier. In Gary's case, for example, it manifested in difficulty with the paperwork attached to a freshly assumed supervisory role at the ice rink in which he had worked since leaving school, on the basis of his disposition towards bodily performance, and slowly acquired more responsibility, soon forcing him to leave the job for another within the rink:

> [...] they asked me whether I'd like to become departmental supervisor, which is basically you're looking after the floor of the ice-rink, so I did that. And then I didn't do very well with some of the book work, this is where my schooling let me down, down to me or the school whichever, and so I didn't quite – *I came to a stop at that point* [...]. (emphasis added)

Even where financial stumbling blocks have ostensibly withered the persistent inequalities of embodied cultural capital, to use Marx's (1968: 96) phrase, weigh like a nightmare on the brains of the dominated. Caroline (25, nursery nurse, daughter of a routine IT worker who died when she was young and a home carer), for instance, decided to retrain in drama and psychology after working as a nursery nurse for some time upon leaving school, first through a BTec course and then at university. This might seem at first glance a demonstration of the extension of lifelong educational avenues and trajectory change to all, and indeed given that Caroline saw no economic obstacles on this pathway considering it was all 'paid for' by government finances (loans and grants) and at least *believed* that university is now 'more open and it is better and easier to get to' and that 'more people have got the opportunity to go' later in life, revealing the changing perception of the possible among at least some sections of the dominated in the 'knowledge economy', it does signal *bona fide* shifts. Yet once Caroline arrived at university, a former polytechnic close to home, the limits of her hard-won embodied cultural capital soon halted progress. She promptly found that she struggled to 'grasp', 'understand' or 'cope' with the work, that it was 'too much', 'very hard' and 'went straight over my head', so she had to 'give up' with the belief that 'it wasn't for me' – a clear indication of the sense of limits inscribed in the habitus – and that, in line with the practical mastery attained at school, she 'should've done more of a practical course', eventually falling back into nursery nursing and thus her previous trajectory with the pernicious corollary that she

> felt a failure, and that I couldn't achieve what I wanted to do. That's what I really wanted to do, and I couldn't do it, *I wasn't clever enough.*
> (emphasis added)

There is, however, another form of generic embodied capital chiefly detectable among the select few interviewees who are experiencing or have experienced upwards movement away from their natal origins in the dominated sector of social space subsequent to their initial entry into the labour market. This is embodied *social capital*, produced when recurrent interaction with a consociate from a higher reach of social space, with their own orientations and perceptions of the possible, within the routine lifeworld feeds into everyday experience and begins to alter and shape expectations of what is achievable. So Gina, a poorly paid learning support assistant who has recently embarked upon a degree course with dreams of being a teacher, admits:

> It's probably through my head teacher, she's the one who pushed me to go on and do things cos she can see that I was more than what I've been doing. I wasn't looking outside the box almost, I thought that

was it for me, and she just pushed me gradually. Other members of staff as well, but mainly the head teacher.

Thus the perception stemming from the classed habitus of what is 'for me' is gradually worn down by the flow of expectations of the possible emanating from an agent whose perception is attuned to rather different conditions (which implicitly denigrates Gina's position with the statement that she is 'more than' what she is), prompting the tentative uptake of new opportunities for lifelong learning. Yet the double qualification forwarded for the upwardly mobile in the last chapter must be mobilized here once again. On the one hand, the fact that Gina – and Martin, whose trajectory mirrors hers significantly but in the field of auxiliary medicine rather than education – is not only in a workplace laden with daily interaction with professionals but *originates from a slightly higher position in the dominated sector of social space* (her mother managed a small shop) increases the chances of *possessing* such contacts and, because her schemes of perception are closer to those transmitting the aspirations, being *receptive* to modification. On the other hand, the invidious and inescapable certainty is that embodied social capital fosters soaring expectations that can only be realized with considerable *toil* against the remainder of one's resources – Gina's original aspirations were not, after all, formulated *in vacuo* but in adaptation to the material and cultural realities of her life-world, now brought to consciousness by her actions as she has not only found the course and its readings 'very difficult' but she and her partner are both 'worried about the money' and have had to labour to secure funding, which has been 'frustrating', 'tiring' and prevented her from doing as well on her course as she could have done otherwise. As if that were not enough, such struggle often results in a *devalued* outcome – in Gina's case, attending a college-based course run by a new university will produce an accomplishment destined to low symbolic worth in the market, if not, as in the case of Caroline seen above, failure.[2]

Interestingly, however, while embodied capital in its generic forms may effectively lock agents within certain zones of social space, this is not, in concrete practice, the typical way in which resources inhering in personal properties guide consideration of prospective possibles when a change of work is on the cards. Instead, as embodied capital is incarnated in any one biography as particular *job- or industry-specific*, if still classed and gendered, skills and knowledge complexes gained from particular training and occupations, its effects are generally realized through the pull of the interviewees' accumulation of *specific* occupational experiences as they meet the structure of available positions provided by the labour market, however volatile the latter may be. In other words, if generic capital sets the broad boundaries of the field of possibles objectively open and structuring the pre-reflexive impression of the possible, practicable and desirable in any particular deliberation, then the specific, occupation-based cognitive and corporeal possessions giving concrete form to that capital tighten the range even further and attach weights to the options within it – something that is at complete odds with Bauman's (2002: 193) rather facetious avouchment that individual skills and 'habits' (or dispositions) melt away with the vagaries of flexible employment contracts and that agents are somehow extricated from 'the legacy of their past' (Bauman, 2004b: 116).

To say that this occupational effect characterizes the work histories of *almost every one of the interviewees*, whether dominant or dominated, incremental manoeuvre or seemingly disjunctive metamorphosis, is no exaggeration.[3] For the most part it manifests in the endeavour on behalf of the interviewees to stay within the same universe of work through job changes, transferring knowledge and skills from one role to the next with minimum modification. Sometimes, namely in the case of the dominant, this takes the form of a steady progression through the ranks of a specific profession in line with ingrained expectations of a 'career', even if, as Savage (2000: 140) argues, nowadays this incorporates not just a change of employer but of employment status (self-employed, consultancy, entrepreneurship and so on). So, for example, we have Nigel, who considers his movement from lecturer to Reader to Head of Department at his university 'a normal part of the academic career', but also Barry, who claims to have made a 'slow progression' from one level of authority and responsibility to the next within knowledge management despite working for different employers and moving from a permanent to consultancy basis, and Courtney, who has worked for a variety or organizations promoting independent film in the UK but gained responsibility (and growing remuneration) with each move, saying of her first job with the British Film Institute:

> [It] has been a good grounding for everything I've done since. It was because of that job that I've been able to get the next job and the next job and the next job.

Projecting this past into the future, furthermore, she adds:

> Film is my background, that's what I've studied and where my path's taken me, so I'd like to stay within that sphere.

Indeed, if we look a little closer at some of those considering retraining we find that the pull of the professional past soon enforces practicality in to the equation. After Nancy announced her willingness to attempt a new venture, for instance, the interview proceeded thus:

> WA: Is there anything particular that's caught your imagination?
>
> Nancy: Well obviously *having worked in planning* I'm very interested in property development from the legal aspect. So that's something that I might potentially consider doing, something in property, and *quite a few barristers do end up doing that* [...] but I think to retrain having become highly specialized would be *a bridge too far.* (all emphases added)

This is not the reinvention or Nietzschean creative destruction that Bauman talks of, but a perpetuation of the past – in the form of knowledge complexes gained through occupational experience – into the future that discounts, as unpractical and, therefore, improbable, the trading of one familiar universe with an alien other. Likewise, Liam (25, economic development officer, son of FE lecturers) admits that his vaguely envisioned fantasy future as a teacher is a 'daydream', and projects the more likely reality that he will gain further qualifications in town planning and continue his career with the local council, while Craig, after announcing his desire to retrain as an architect, soon concedes that such an aspiration is not so simple 'when you get down to it' and that, therefore, 'I probably won't change [...] I really don't think I'm going to become an architect'. Even Sean's decision to retrain as a software developer only makes sense when situated in his trajectory: having slowly inched from photography over to the digital archiving of photographs for newspapers, he had already begun to pick up specialist IT skills, which can be considered a form of specific cultural capital insofar as it requires a level of symbolic mastery (Kapitzke, 2000) – it was just a question of augmenting them.

Among the dominated the inclination to remain within the same broad sphere of work is just as evident and steered by the drag of occupationally specified capital – usually in the form of what Bourdieu (2005) called 'technical capital', or vocational skills, acquired through schooling and an apprenticeship, which allows circumscribed advantages within the local market of lower social space. There are, however, two significant dissimilarities with their privileged counterparts. Firstly, progression through a 'career' is generally substituted for horizontal or even, occasionally, downwards occupational movement through the life course. This is the case for Joe and Dave, for instance, who, prompted by the unforeseen disintegration of their employment witnessed earlier, opted for occupations offering the same or lower returns – building maintenance and long-haul lorry driving respectively – for which they already had declarative and procedural knowledge (and credentials) because of the trajectories they had travelled since school, while Gary's current consideration of a lateral move from driver instructing to driver examining is based on the fact that the same constellation of skills is deployed. Though trade apprenticeships and the acquisition of technical capital, as the credentialization of physical practical mastery in line with the construction of valued masculinity (necessarily valorizing what they have) among the dominant, still remain largely the preserve of men in the UK, the female interviewees among the dominated echoed this process. As Gillian, who had passed through a series of low-level office jobs doing 'a similar sort of thing', put it, encapsulating the bounds of consciousness in the transition process, here brought to awareness only by prompted reflection on her change away from clerical work: 'Then you're in it [your first job] and that's it, that's all you can do and so your next job's going to be that, and the next one and the next one.'

Secondly, where the dominated obtain promotion to supervisory roles and lower management with their employers, and with it slight shifts in social space, it assumes a fundamentally different character to the ascent of the dominant. In contrast to the taken-for-granted progression of a more or less defined 'career' seen above, movement for the dominated – who, interestingly, justified their actions with a discourse of 'moving on' and 'bettering oneself' and thus become the upper-dominated parents witnessed in the last chapter when discussing the upwardly mobile – is usually *piecemeal* and often undertaken with a certain amount of trepidation. Gina, reflecting on her continued journey of retraining described above, puts it most clearly, but Jimmy is also revealing in his discussion of his future in the context of his recent flirtation with supervision:

> It was *always small steps*, I've *never planned anything too big*. It was 'I'm just going to do my GCSEs', and improve them. Once I did that – in fact that's the whole way my life is, all the courses, it's just, I don't think 'I'm gonna' – but I would like to be a teacher, so I suppose I do have a goal now. But it's always been, I'll just do my GCSEs again. Done that, okay, let's do car mechanics course and that, then I got the job, and so it was 'I should do a teaching assistant's course', do that.
>
> (Gina, emphasis added)

I suppose like anything, you'll go one level and then you'll look at it, the next level, and see if it's – I'll just play it by ear [...] I'm just happy to carry on as I am, *taking one step at a time* and just looking [...] At the moment I'm just happy to move along at a steady pace.

(Jimmy, emphasis added)

Such caution and anxiety appears to stem from two sources: fear of failure, based on a perception of the limits of one's embodied capital, and fear of dislocation from the workplace culture, centred on the dominated ethic of 'having a laugh', 'staying together' and 'us versus them', specified by the specific people and places of the lifeworld and constitutive of comfortable familiarity (cf. Beynon, 1984; Hebson, 2009). Once again Gina expresses the former lucidly in rounding out the above passage, while Jimmy displays the self-questioning and anxieties that stem from the latter:

So it's always not been high. I've never set anything too high, because I don't like failing and I don't think you can plan your whole life out.

(Gina, all emphasis added)

WA: Have you had any issue going into management in terms of how you're seen by your workmates?

Jimmy: Yeah, definitely. Obviously some didn't like it, some didn't care, but you definitely get a – within myself, *sometimes you question yourself*, but I presume that's horses for courses, it happens to everyone. But there was definitely some who shook your hand and said well done, 'everyone should have a go', others were like 'you're going to the dark side', things like that. (emphasis added)

These reservations can, in fact, arrest advancement altogether, as when Maureen decided not to take the plunge into supervision on account of an aversion to 'responsibility' and a desire to 'get on with everybody' as 'one of the lads', but equally it should be acknowledged that they are not always an issue. Some of the dominated, males in particular, had no problem at all with the prospect of striving for or taking on supervisory positions, pursuing them with the belief that 'I can do that' (Dave), but it is revealing that these interviewees – Dave, Phil, Chris and Gary are prominent examples – had parents in supervisory or skilled occupations. It could well be, therefore, that they originate from a higher stratum of the dominated section of social space that fosters the desire to 'progress' and 'move on' but, lacking the level of resources (hidden or not) or focus on education necessary to ensure upward mobility through the education system among the highest stratum of the dominated, demands it be realized through working life instead.

This is not to say there are no interviewees experiencing disjunctive movements, where trajectories involve stepping out of one universe and into another thanks to changing employment practices or opportunities – we saw plenty of these at the outset of the chapter. Yet even where this occurs the same logic of specified embodied capital underpins each move. For the dominant, such ruptures are usually horizontal shifts in an otherwise steady ascent and remain bound by the classed occupational past as it intersects with the structure of available positions. Looking in some detail at two cases that, on the surface, are exemplars of 'de-standardization' should serve to demonstrate this.

Helen, first of all, started out her post-university life as a special needs teacher, only to then move into international sales with a company designing and selling children's play spaces and then, after time out of the labour market for children, marketing for a business specializing in resettling workers after redundancy (the existence of which is itself telling). By all accounts an erratic series of unconnected vocations veering from the left to the right of social space that, were Beck to glimpse it, would be rubber-stamped as reflexivity in action. If, however, we dig a little deeper into the narration of these transitions we can detect the pull of the complexes of knowledge that, as specific embodied capital, contribute to the constitution of the habitus. Firstly, when Helen departed university with her PGCE qualification in hand her projected future was as a teacher in a mainstream secondary school but, finding work hard to come by given the demands of the labour market, had to be satisfied with the compensation strategy of teaching in the aforementioned special needs unit. Falling below her threshold of expected economic remuneration given the anticipations fostered by her background and training, however, she continued searching for school-based employment, in the course of which, when perusing the newspaper, she happened upon an advert for a job with her next employer, which, she says 'was still working with young people', 'still [had] a lot of educational elements in it' and ultimately 'mentioned it was an educational play company in the advert, *so I suppose that's what drew me to it'*. In other words, her practical sense of what was viable given her accumulated, occupation-specific (but also, because a form of cultural capital, class-based) stock of knowledge, in line with the 'principle of pertinence' (Bourdieu, 1990a: 90), instantly and pre-predicatively guided her selective attention towards this avenue, placed it within her field of possibles and prompted action.

After acquiring the job, she expected to return to teaching before long, but soon excelled at sales and received promotions, including taking on more and more managerial roles, largely, she says, because *'I'd always been quite commercially aware anyway, cos dad had obviously been in sales* and I'd always been quite interested in commerce' (emphasis added). Knowing that her father had been the sales director for a successful company throughout her childhood, and that this, as she herself admits, fed into her orientations, abilities and 'awareness', allows a new insight: far from veering almost randomly around social space with the reflexive impulse, she started out from further towards the right of social space and simply took a detour through teaching before falling back on already acquired dispositions and reproducing her familial position in social space, demonstrating, unfortunately for Archer, the disproportionate pull of early class-based lifeworld experiences. And what of her subsequent move into worker relocation? Again, the same logic persists: when returning to work through agencies after a considerable period out of the workplace to have children, they placed her with her current employer in an office management role on the basis of her previous experience of such tasks in the play-spaces job (once more matching past acquisitions with the structure of available spaces), but when the marketing director left unexpectedly she had to temporarily fill her shoes until corporate restructuring blended the two roles anyway. Now, having been submerged in it for some time, she is considering further study in marketing to further increase her pay and 'prospects', thus resuming upwards ascent in her career, stating that she wishes to perpetuate her current skill set because she now has a 'background in it'. Future 'radical' ruptures, she adds, are strictly forbidden.

The second case is Paul, the son of a builder and a secretary, who detested school and is, thus, one of the few upwardly mobile interviewees not to have attended university. Having worked into management positions in the waste and transport industries, he was unceremoniously dislodged from his position in the face of 'downscaling', only to turn his hand to painting and decorating for a period and then, seemingly from nowhere, to software development. He himself was adamant that the idea of having a 'job for life' 'doesn't exist anymore', that 'people get fed up, people need a change, people grow out of what they're doing I guess' – certainly, he does not 'really want to do the same thing all the time', and, echoing the *faux* reflexivity of others already witnessed above, used his redundancy as an opportunity to 'think about what I wanted to do'. But even in this trajectory, which would otherwise appear to have broken from class and occupational bounds in every conceivable way, previous experiences 'stake their claim to eternity', as Merleau-Ponty (2002: 457) says. In particular, Paul's trajectory is patently based on an oscillation between two facets of his upwardly mobile habitus when faced with the forces of the economic structures of society. On the one hand, lacking many of the capital stocks open to the university-educated, when he was made redundant and became a self-employed painter and decorator he fell back on the specific *practical* skills that he had first gained ample experience of in his dominated early lifeworld working with his grandfather, a builder and property renovator. On the other hand, when, finding that financially unworkable, he opted, like Helen, for the options provided by employment agencies, he was plugged into software development (which, though amply remunerated, is separable from programming in that it requires fewer skills) *because he had already attained previous embodied (and objectified) cultural capital pertaining to computers* in his managerial jobs, and he followed willingly because this prior experience had, as he put it, 'caught his imagination'. Yet the relative strength and primacy of his earlier experiences upon his current position, as with Helen, is significant – he intends only to stay in his present line of work for a few more years before eventually moving into property development, a lucrative application of his renovation skills – even if he will need the economic capital he is accumulating now, as a well-paid 'wired worker' with an equally well-paid human relations manageress for a wife and no children, to achieve this.

What Paul and Helen both show, then, is that even where the kind of processes said to be on the increase in neo-liberal knowledge societies – redundancy, restructuring, workplace learning and, in the case of Helen, a mismatch of credentials and available positions – have altered the structure and dynamism of the labour market (and let us not forget the evidenced calls for moderation marshalled at the outset), it still meets with the inertia of classed *Erfahrung* deposited in skills, abilities and propensities, at most impelling, in the case of the dominant, steadily ascending careers and social reproduction to traverse a few occupational universes on their way. The same is equally true of those *dominated* interviewees experiencing diverse occupational shifts, albeit without the continual climb. Here, however, the underlying specified dispositions manifesting embodied capital evince, more than anything else, adaptations to a *lack* of valued resources. Take, for example, Eddie, who, as we have already seen, has journeyed through a string of unrelated jobs – lifeguard, motorbike salesman, order picker, fire fighter, caretaker and, he hopes, having applied for the position, school groundskeeper, with the added possibility of sports coaching. A diverse concatenation, for sure, but nevertheless the guiding thread is a habitus, a life, dedicated to two practices – sometimes conjoined, sometimes separate – which, as we have seen, are common products of the practical, bodily mastery developed in school and valued through life when cultural capital is in short supply: sports (and, by extension, physical activity) and 'the outdoors'. Witness:

> I was always into sport, always active. I always knew from a very young age I wanted to do something linked with sport, whether PE teacher or being a PTA in the army. That type of thing always interested me. If I was going into the army, doing the assault courses and that, that always interested me. Being outdoors, doing outdoor activities, if I hadn't have gone into that I would've done an Outward Bound course or something like that and gone into mountain leader training. Anything to do with sport or the outdoors, no matter what. I did an athletics training course through work and that's probably out of date nowadays, quite a few years back, but from a very young age I was always interested in and wanted to do something related to sport. [...] *I know sport, but that's it really. Anything linked with that, I'm quite happy.* (emphasis added)

Lifeguard, fire fighter, groundskeeper (which includes tending the sports field and equipment) – the pursuance of all (and other jobs unsuccessfully applied for, such as park ranger) are dictated by the perception of them as potential avenues for activating in some way his knowledge of and interest in sports, physical activity and outdoor life. Thus, despite realizing that PE teaching was beyond his grasp, Eddie defiantly declares:

> I still got into sports, with the lifeguarding, and I was an instructor with scouts and cubs. So I was doing sports and outdoor activities all the time anyway. I'd go to work, get some money, then do the outdoor activities.

Similarly, he wasted little time in applying for the groundskeeper's job in order to leave behind his caretaking work at the same institution, taken as a stop-gap to meet economic exigencies when made redundant from the fire service, once the opportunity (on the previous incumbent's retirement) arose: 'I enjoy sports, and I enjoy outdoor life, so working as a groundsman would be pretty good.' Fire fighting, with its appeals to physical ability and masculinity, was also seized upon when offered, while the job selling motorbikes seemed to harmonize, at first, with his long-held interest in such vehicles as ways of exploring the outdoors. However, when it transpired that selling motorbikes is not quite the same as riding them, his dispositions, in conjunction with economic necessity, propelled him to leave: '[It wasn't] the best of jobs, cos I'm an outdoor, moving about type of person, not sat behind a desk.' Similarly, we saw above that Eddie was not 'happy' with his order-picking job and 'wanted out', *and now we know why*: it was, he says with distaste, 'indoors all day'.

This dual disposition driving Eddie's work life is not reflexively formulated, knowingly and playfully constructed from encountered bits and pieces, as Giddens would have it, but anchored deep in the classed experiences of his youth. Sport is, as we saw in the last chapter, a common avenue for the realization of practical mastery, and this is certainly the case for Eddie: he was not very 'academic' because of his parental capital levels, and did not like

> Sitting down and reading and stuff, yeah. Never into reading. [...] I was always the one out, messing around [...] At the time I wasn't really into learning as such, *was more interested in going out and playing and swinging from trees and stuff like that* [...] We'd always be playing football in the garden. He [my brother] was quite into sport as well, and we'd always be racing each other, climbing trees and always being very active since we were little kids. (emphasis added)

A reaction to a lack of cultural capital, but also the product, it seems, of the *laissez-faire* parenting (Eddie's dad was not 'too involved' and never 'pushed' or 'encouraged' him) uncovered by Lareau (2003) and already witnessed in the previous chapter among dominated parents. Based on the fact that the parents do not posses, like the dominant, the knowledge of how to or belief that they can push their children to succeed in the school, they nurture a split between the world of adults and children and a faith in 'natural growth', which results in youngsters like Eddie and his brother, rather than being stuffed full with valued resources through focused activities, finding their own disposition-building entertainment – in their case, in sport and the outdoors activities available in their semi-rural lifeworlds.[4]

Eddie's trajectory may have revolved around sports and outdoor life, but this is not the only way in which a classed disposition rooted in the absence of embodied capital can motor employment histories. Andy's moves from graphic designer to cartoonist (for a music magazine), nightclub DJ, screen printer (specializing in printing music posters) and now supervisor of a studio shooting music videos, for instance, often prompted by the insecurities characteristic of the kind of work open to those in the lower reaches of social space, have all been underpinned by a commitment to and knowledge of specific popular (and hence dominated) musical forms (especially punk, hip-hop and dance) as a part of a package of politicized symbolic resistance to the deprivations he and others in his position have faced, first initiated in and carried forth from adolescence, not dissimilar to that seen among Andy's generation by Hall and Jefferson (1978) and Hebdige (1979).

Conclusion

The overall message is, once again, bad news for the reflexivity theorists and confirms many of the reservations and counters forwarded when reformulating their claims as feasible hypotheses. This is not to say there have been no changes since the days of embourgeoisement: shifts in the global economy and employment policy in the post-war period, and especially with the rise of neo-liberal governments in the West partial to deregulation, human capital and the destruction of stability, have fostered greater insecurity – in perception at any rate – in the labour market and augmented options for retraining and perhaps even the need for more 'choice' and 'decision-making' throughout one's working life than in the past, at least for the disadvantaged minority experiencing truly de-standardized 'careers'. But, just as in the last chapter, these changes in the social landscape should be seen as providing nothing more than the shifting *context* of class; the structural *relations* of difference and distance that constitute social class, as we have seen, continue to differentiate employment histories. Those with ample capital, on the one hand, are either committed to secure, materially and symbolically rewarded careers or at least possess the freedom, because of their resources, to exploit the new options and opportunities and handle episodes of enforced change, whereas the dominated, on the other, are guided by the material pressures, limited capital and classed dispositions inscribed in their situation. The un-socialized reflexivity posited by the reflexivity theorists is absent, and in its place we find consideration of the field of possibles in mundane consciousness, liable to be misread as reflexivity, and that while the precise nature of what appears within that field – retraining, multiple jobs or whatever – might well be novel, the relative spread of social space it covers is, by and large, not.

6
Distinction and Denigration

The idea that class has ceased to structure life courses was always the more controversial and contentious component of the reflexivity thesis. Much more credible, so most commentators admit, are the claims that class lifestyles, differentiation, explicit identities, talk of class and class politics – in short, the various elements of the *symbolic* dimension of class – have waned to the point of extinction and been supplanted by atomization, personal responsibilization, reflexively adopted leisure interests, subcultural or other affiliations and post-materialist politics. But, leaving politics and explicit class discourse until the next chapter and focusing here on lifestyles and social identity, some prefatory reflection is necessary to ensure conceptual rigour. Specifically, what precisely would it take for a phenomeno-Bourdieusian model of the structuring of lifestyles and identities by class to be defeated by social change? If globalization and affluence have opened out the lifeworlds of the population to a spectacular array of new and exotic practices while relegating others to history, does this *ipso facto* demolish class lifestyles and identification?

Bearing in mind the fact that Bourdieu's is a relational rather than substantialist vision of class, the answer is clearly no. In terms of life-styles, it matters not what the concrete practices and products are so long as there are broad homologies between practices in the symbolic space and positions in social space that conform to the causal logic of adaptation to material and cultural conditions of existence mediated and individuated by lifeworld particularities. Furthermore, it is not nec-essarily the practice alone, but, especially where practices seem to span the space of lifestyles, *how* that practice is perceived, constructed and conducted that matters, such as when the members of the dominant class apply an aesthetic honed in distance from necessity to a practice ordinarily perceived as 'common' (Bourdieu, 1984: 40; see also Holt, 1997; Rimmer, forthcoming). On both these fronts there are serious blows for those reflexivity theorists like Giddens and Beck who lay special emphasis on the uncoupling of lifestyles from class delivered by the research of Bennett et al. (2009). Homologies abound, though, they argue, it is less about legitimate versus popular culture, as it was for Bourdieu in *Distinction*, than it is about the capital-rich displaying a taste for a wide and eclectic range of practices and goods from a range of genres, from polo to pool, Mozart to Madonna (what Richard Peterson [1992] famously dubbed 'omnivorousness'), versus the capital-poor who tend to stick to one type of practice or genre (the 'univores'), and the orientations individuals can have from different regions of social space towards the same practice can vary widely. However, as we already know, there remains little examination of the *genesis* of taste in this otherwise ambitious research project, meaning that reflexivity is let off the hook and still in need of examination. In fact, as the first part of this chapter will endeavour to show, when we do descend from the quantitative level, not only does the notion of reflexivity find little support, but the omnivore–univore division is shown to be more complex, perhaps even more spurious, than often supposed.

As to identities, what matters is not whether people explicitly envision themselves as a unit in a greater entity anointed with a class label, as in so many older studies of class identity, but the multiple modes through which people *distinguish themselves relationally from others* – either concrete individuals or specific constructed 'groups' – within the social space, display a keen sense of difference and similarity in narrating their lives and convey this in whatever register they have available – all of which, following Bourdieu, would coincide roughly with the objective divisions of capital and latch onto homologous practices. Others have begun to unveil these processes among specific groups (Skeggs, 1997; Savage, 2000), but the argument in the second part of the chapter builds on these pioneering studies to demonstrate, against the reflexivity thesis, their generality, highlighting the endemic, stark and often self-depreciating *sense* of difference and similarity – or class sense as Bourdieu dubbed it – that orients perception of self and others, even among self-proclaimed doubters.

Class practices

To start with, an almost banal point to comfort the reflexivity theorists: economic and socio-cultural change have had an obvious impact on lifestyles. Observation of the interviewees' living spaces and analysis of their reported pastimes reveal that some consumption practices once linkable to divisions of capital, such as ownership of televisions (including wide-screen models), cars or mobile phones and taking foreign holidays, no longer, in terms of sheer prevalence at least, act as reliable signposts to class differences thanks to affluence and technological advances, while old amusements and one-time hubs of communities, such as the Working Men's Club of Dennis et al.'s (1969) coalminers, have dwindled and new pursuits, products and activities, whether indigenous or – like Thai boxing (Liam), jujitsu (Gina), Salsa dancing (Sue), jive dancing (Yvonne) or Japanese sushi (Hannah) – appropriated from distant cultures, have sprung up in their place.

Beyond this modification of the concrete matter of lifestyle practices, however, few empirical stanchions can be uncovered to sustain Beck and the others' theses. Instead, though the qualitative nature of the research forecloses confident conclusions on national patterns, the symbolic practices and tastes of the respondents are, by and large, clearly divided by the enduring social partition that runs between the dominant and dominated and their uptake fully explicable with a phenomeno-Bourdieusian theory of class. This is, as will now be demonstrated, as true of the cluster of activities practised in the private seclusion of the domestic sphere as it is of leisure jaunts in that domain usually targeted by concerned inquiries into 'cultural participation' premised on a narrow view of 'social action' (see, for example, Chan and Goldthorpe, 2007b: 3), the public sphere.

Class at home

Peace in the home may bring happiness to the king and peasant alike, as Goethe is supposed to have said, but when it comes to specific home-based pleasure-producing activities universals uniting the socially distant are harder to come by. Interviewees populating the higher region of social space, for instance, frequently forwarded three activities more or less absent among the dominated: gardening, cooking and DIY.[1] The first of these, the 'purest of human pleasures' according to Bacon, is not simply a product of possession of a garden versus dispossession, though that plays its part, for plenty of the dominated have gardens as well. Equally important, it seems, is the *orientation* to the garden and its upkeep generated by material conditions. It was noticeable that the dominant respondents' gardens were usually neat, well-tended and focused on plant varieties, with the activity itself described as a leisurely 'picking things and trimming, moving pots' that takes 'a huge amount of time' if it is to be done to required standards (Jackie), all of which indicates an *aestheticization* of the garden generated by the possession of temporal and economic freedom. Compare this with the diminutive and unkempt outdoor spaces of the dominated, strewn with garden furniture, old motorcycles and the like, and, if they had children, toys and animal hutches, which appeared to serve the *practical* function of storage space and arena for youngsters to gambol unsupervised and whose tending involves more fundamental activities like clearing up, cutting the grass and so on that are more likely to be perceived as tiring chores than sources of pleasure.[2]

The same is true of cooking – generally a female pastime, demonstrating the persisting gendered division of labour in the household through the social space, but not without its male enthusiasts. The dominant interviewees told of relishing the testing of new recipes, doing something 'different' to express 'your own sort of style' (Oliver), having 'a love of food' (Martin) or taking pleasure from 'good food' (Tessa), usually in the company of friends in the form of a dinner party or 'allotment party' (Oliver) as much as once a week (Abby), all of which, as with gardening (and DIY), indicates an elevation to the status of self-realizing *leisure* pursuit that which for the dominated, with fewer resources, depleted time and hungry mouths to feed, are *functional* activities. The dominant have the time, the space, the freedom and the distance from necessity to playfully experiment, invest energy and bear the consequences of failure, and so, separated from the urgencies of material exigency, they develop the 'pure aesthetic', which constitutes as an object of expression and aesthetic delight that which for others is inseparable from the practicalities of subsistence (cf. Bourdieu, 1984: 40).

The dominated, by contrast, listed fewer home-based leisure activities – indeed fewer recreational practices at all – largely because, owing to the nature of their work, time, energy and money are in shorter supply. As Dave responded, he lacks 'much spare time' and certainly 'the time or the money for anything expensive' (Dave), while Caroline laments that she is 'too tired' to have a 'social life' after the long hours caring for the privileged children in the nursery that employs her, and that 'It's just finding the time, I don't have enough time to do other things, see everybody'. Indeed, because restricted time and economic capital preclude countless pursuits, home-based activities billed as 'leisure' sometimes, as in Eddie's case, include the snatched 'family oriented' moments that typically evade questionnaires. As he puts it:

> [Because] I haven't had the greater pay jobs, we utilize our free time more and spend it together as a family doing family activities. If it's just playing out in the garden or going for a bike ride on a Sunday or something, using what time we've got to greater advantage.

Nevertheless, some of the dominated interviewees did mention discrete activities undertaken in the home and its environs, including 'making things', such as craft models or garments (Tina), repairing and servicing friends' vehicles (Gina), and 'tinkering' with motorbikes (Dave) – activities centred more on practical or functional endeavours entailing the bodily craft and skill comprising practical mastery. What is more, the specific provenance of these concrete activities can sometimes be traced and, far from unmasking an antecedentless, reflexive adoption, as Giddens would have it, demonstrate the hold of particular capital-mediated childhood experiences, classed dispositions or parental input. For example, Dave's penchant for 'tinkering' with motorbikes – which he admits has waned now he is older – is contextualized by the early lifeworld experiences granted by a father who had worked as a mechanic and whose labour spilled over into his spare time:

> Dave: My dad was very much like you know, he was very keen on vehicles and stuff like that in general and *we kind of grew up you know tinkering around with cars* and what have you, and I suppose it's where it comes from really, no doubt about it.
>
> WA: So what kinds of things was he into with his cars?
>
> Dave: He'd just, you know, take the engines apart, repair them, you know whatever, always mucking about with them in one way or another you know. Yeah, and you know when I used to have motorbikes I used to do that as well, used to take 'em apart and mend 'em you know, put 'em back together again. (emphasis added)

There are, of course, some home-based practices that dominant and dominated alike indulge in, but even here, once we venture beyond the simple listing of activities, each practice encloses an intricate universe of fine class differences. In some instances inadequacies of the data mean we can only surmise the possibilities. Computer games and watching television, for example, are both evidently popular pastimes from top to bottom of the social space (the former especially among younger interviewees), but if Bennett et al.'s (2009: 132–51) research is anything to go by they may well be differentiated in frequency and content. In other cases, however, we are well equipped to dissect the internal complexity of practices apparently uniting disparate areas of social space and, indeed, which Bennett et al. failed to provide any unambiguous class-based conclusions for. This is nowhere more the case than with *reading*, a practice reported by many of the respondents from across the spectrum of social space, for what is interesting here is not so much that the dominant were slightly more likely to forward it as a leisure pursuit or be enthusiastic about it, though that is already telling, but the differentiation of the *types* of book read.

On the one hand, the dominated read only three varieties of book: biographies (usually of 'celebrities' of one kind or another, such as David Beckham, Freddie Mercury, Radio One presenter Chris Moyles, and – he could hardly be described as anything other in contemporary Western society – Che Guevara), popular fiction including Dan Brown, J. K. Rowling and a gamut of romance, crime and spy thrillers, and occasionally non-fiction on, for example, diets and 'self-help' (Maureen) – all forms of writing that, in topic and language, revolve around the *concrete experience of the (fictional or real) individual* and the *practical business* of living, feeling and deciding in 'real' life rather than the abstract issues beyond and underlying the interpersonal world of interaction.

Compare this with the dominant who, instead, overwhelmingly study material that engages their symbolic mastery, their *propensity for abstraction*, which comes with the possession of cultural capital, with the specific topic through which this is realized being differentiated according to lifeworld experiences such as those of work or travel.[3] Thus they read about national and international political and social matters or current affairs (*Tescopoly, Fast Food Nation, Watching the English*, books on Africa and on Tony Blair were variously forwarded by Liam, Karen and Debbie, for example), about history (in Martin's case, with a peculiarly narrow interest in 'Vespasian and the Flavian dynasty' of the Roman Empire) or indeed anything 'vaguely sort of informative' such as Richard Dawkins' *The God Delusion* (Lisa).[4] The underlying disposition reveals itself in Courtney's account:

> I like travel books. *I've learnt a lot of my general knowledge or history of the world comes from novels*. So books about people in different places or lifestyle books set in different times and places. I don't really remember that much from history or geography from school, so I kind of feel I've got a lot of my history knowledge through reading. And because I've travelled quite a bit I like to read books about the countries I've travelled to or read something about that country while I was there so *it put things into context*. (emphasis added)

Using novels and non-fiction books to acquire and expand historical and geographical knowledge and to 'put things in context', that is, to be disposed to situate one's concrete experience (in this case revolving around travel) in the abstract system of world history – sure indicators of the motivational push of symbolic mastery in the use of the written word. And then, of course, there is the propensity among a notably large number of the dominant to read the 'classics', as they themselves entitled them, including Dante, Homer, Ovid, Flaubert, Tolstoy, Dostoevsky and Herman Melville, or else otherwise symbolically valued and consecrated (through, for example, Nobel prizes or Booker prize shortlists) latter-day authors such as Umberto Eco (Diane), Margaret Atwood (Abby) or Mohsin Hamid (Nancy), all of whom are credited as literary dissectors of phenomena transcending the conduct of practical life.

Often yoked to this was an expressed *distaste* for certain types of book – with the fare of the dominated, biographies and certain forms of popular fiction, coming in for especial scorn – on the basis of an aesthetic that values the mobilization of symbolic mastery. Thus Diane describes the written corpus of the glamour model Jordan as 'rubbish books which I wouldn't even dream of having in my house', and Rebecca says of biographies that 'It doesn't really feel like you're reading a book, it's just someone writing about themselves', but perhaps Nancy expresses it best:

> I'm not keen on autobiographies and biographies and factual based books. I like a novel, a fictional novel that's *well-written and high quality*. I don't like bodice-rippers or girly books. I like something that's *well-written, thought provoking and a good story*. (emphasis added)

In other words, via the medium of education and cultural capital, objective distance from practical necessity has produced a system of perceptual schemata geared towards *mental* distance from practical reality, encapsulated in a demand for 'quality' writing, characterized by syntactical and lexical complexity and accuracy, and stimulation of *leisured rumination*, which necessarily entails a rejection, as inherently inferior, of 'bodice-rippers' and 'girly books', that is, human interest stories rooted in visceral urges.

Yet it would be a mistake to suppose that the dominant were uniformly univorous in their quest for abstraction, for, in nearly all cases, there seemed to be an interspersing of consecrated literature and non-fiction with popular fiction and biographies. This was done, however, with the self-conscious justification that they occasionally seek 'light entertainment' (Liam), 'escapism' (Debbie) and to 'switch off' and not have to 'think about much' (Sonia), explicitly describing such books as 'trashy' (Adrian, Sonia), as if a sort of gentle sparring is appreciated between more challenging tomes to relax and refresh (cf. Lahire, 2004: chap. 16;

2008: 175–6). Thus, and insofar as they still prioritize (by naming first) the legitimate and non-fiction works and perceive the popular books as inferior, while this mixing may be described as 'omnivorousness' it is only of a markedly limited variety and still follows the explanatory logic of a theoretical system devised long before that idea became common academic currency.

Exclusive excursions

So much for the expression of lifestyle differences in the private sphere. But what about the activities propelling agents out of their abodes and along time–space paths in the public arena? Two broad species of activity were mentioned in this regard, one by both the dominant and dominated and another by the dominant alone. These are, respectively, playing (or even coaching) sport and recurrent expeditions to the ensemble of institutions showcasing legitimate culture: museums, art galleries and the theatre.

If we begin first of all with sport, we find, as with reading, that most of the interviewees mentioned playing some form of sport as a typical form of amusement, but that the *strain* of sport practised divaricates along the fault lines of social space. The dominant, on the one hand, listed tennis, squash, rowing, cricket and rugby most frequently, all activities that – in terms of participants (beyond school level) but also spectator numbers across the nation – are generally *rarer* and therefore, one might surmise, more *distinguished*, on account of one or more of the following characteristics. Firstly, effective participation necessitates the possession of expensive equipment (racquets, unisuits, waterproofs and so on), not to mention regular training and practice sessions, which can often, as in Tina's case, dissuade those with less economic capital:

> I was going to start playing squash, mainly to keep fit, only down the road. But you've gotta buy your own racquet. What do you mean buy me own racquet? I just wanna hire one, it's easier. And I might not like it, I might get donked on the head with the ball.

Secondly, and especially so for the team-based varieties, there has long been an association between the sports mentioned by the dominant and the bourgeois public school and university system of Britain, where resources and available time for competitive sports are ample, meaning that the dominant are more likely to encounter readily available opportunities for uptake – and indeed, this is where many of them first participated or became interested in their chosen sport. Thirdly, and entwined with this institutional embeddedness, the above-mentioned sports are, in British culture, bound up with a classed and gendered history and contemporary construction as 'gentlemanly' activities emphasizing 'fair play', respect, humility and self-constraint or, in a word, *civility*, harmonizing with a specific repertoire of dispositions, as opposed to the *vulgar* (or the 'crass'), the *barbaric* and *bad manners* which, as Martin claimed, set the mass activity of football apart from rugby:

> Football I can't bear, can't stand it. [...] If it hurts that much, get stretchered off and stay off instead of rolling around on the floor. Get back up and run off and continue playing. The dissent within football is teeth-grindingly infuriating. *The disrespect shown to other players, the referee, all of that, the obscene amount of money they earn.* The fact that these skilful but *ignorant* people can earn so much money and spend it on such *crass bling*, is just incredible. (emphasis added)

By contrast, precisely this popular disport, football, was cited most frequently by the dominated as the principal sport to be practised (including in a coaching capacity) in time away from work, though others, including some long associated with the dominated such as skittles as well as newer or imported practices like pool and boxercise, were also represented. Importantly, in each case we see not the reflexive adoption of the practices in line with an identity constructed in a milieu of multiple options, but the tacit absorption of dispositions towards the activity through the combination of classed conditions of existence and lifeworld particularities.

Jimmy's passion for football, for example, was forged in a lifeworld that combined, firstly, the unsupervised play characteristic of dominated childhoods (he and his nearby cousins would 'all go off playing', especially football in 'the backstreets', and, 'through winter time' when 'there ain't much else to do with the darker nights and that', he'd attend games with a neighbourhood consociate); secondly, the ubiquitous but unfulfilled dream of converting available bodily skills into material and symbolic success (he was 'thinking wouldn't it be great if I was out on the pitch one day'); and, finally, the taken-for-granted surrounding experiential backcloth emanating especially from his father, who was 'always into football', that 'the beautiful game' is just that and, therefore, should be the primary conduit for his self-made entertainment and improbable fantasies.

Likewise, Tracy claims she took up skittles – a sport with a particularly strong tradition in the South West of Britain – and played in a league for twelve years because

> My mum played, my sister played, my older brother played. So it was just a case of go along, have a laugh at the time, have a game and if you're any good you're good and if you're not you're not, just do your best.

Again the practice infused her lifeworld through multiple family contacts as a 'natural' thing to do, but underlain by the fact that, being a 'pub game' like billiards and darts, skittles is inseparably linked to the propensity to frequent the public house in pursuit of the 'convivial indulgence' favoured by dominated individuals in their free time (Bourdieu, 1984: 179).[5]

As a final example, Phil's penchant for the recently invented activity of boxercise – a form of group exercise that takes boxing technique and fitness as its guide, is heavily meshed with its parent sport (the trainer is a boxer and the group attend boxing events) and, so says Phil, is populated overwhelmingly by 'tradesmen' and, indeed, was first brought to his attention by a colleague at work – is a product of his disposition towards bodily performance and sport developed as a child with little interest in academic activities at school.[6]

There were, however, several physical activities, loosely categorized as sport, named by both dominated and dominant in relatively even numbers, most prominently walking, running and golf. Yet these can be contextualized by, firstly, statistics showing that, though walking is a notably popular activity among the populace, all three activities remain primarily the preserve of the upper ranges of social space (Roberts, 2004: 61), and, secondly, the knowledge that the same practice may take very different *forms* and have divergent *meanings* according to class position (Bourdieu, 1984: 211). 'Walking', for example, denotes, for the dominant, long hikes in the countryside (Exmoor or the Lake District) with full regalia (boots, waterproofs and so on), but for the dominated usually (though not always) refers to more local and sedate outings, such as a Sunday stroll by the nearby canal (as in Maureen's case) or a fair-weather amble with the children (as for Phil). Equally, the orientation towards and desired outcome of 'running' differs for the dominant and dominated – for Rebecca, for instance, it is a means of keeping 'healthy' and 'fit' in line with the ascetic health ethic of fractions of the dominant class where the body is an end in itself; for Joe it is a cultivation of competitive prowess using the body in an instrumental fashion (see Bourdieu, 1984: 208ff; 1993a: 129ff; cf. Savage et al., 1992; Bennett et al., 2009: 159).

If sport and physical activity are a question of subtle differences and divisions, the second collection of activities drawing people from their homes and into the public sphere in the name of leisure is comparatively straightforward. Simply put, a sizeable number of the dominant recounted their enjoyment of theatre (including Shakespeare), museums, foreign cinema, opera and art galleries compared to only a handful of the dominated. Some typical examples:

[at the theatre I like to see] everything. We're going to Shakespeare next week, *Romeo and Juliet*. We saw a play in Bath a few months ago, it was really awful actually, it wasn't very good. In London we used to go to a lot of plays at the National Theatre, so like *The History Boys* – fantastic, really good plays. I like to go to museums, yeah that's about it I guess.

(Rebecca)

[I] really like the theatre, I raise funds for the refurbishment of the Old Vic [theatre] and I'm a patron of the Tobacco Factory [theatre], so I like the Shakespeare. [...] I'm [also] quite interested in going to art exhibitions, certainly used to in London cos I used to go with a girlfriend. If there was an exhibition on we would always toddle off and see it, whatever it was. I'm happy to go and see anything really if it's art, whether it's modern or Impressionist or classical.

(Nancy)

When we go city breaks, things, we'll go to art museums and galleries and I have an appreciation for art. I don't have a great deal of knowledge about art history, but yeah I do enjoy that. That's definitely something – we never go to a major city without going to the major art gallery.

(Samuel)

I've been known to go to the opera quite a bit when I lived in Italy cos it interests me. I'm not sure it's something that I love, but I'm certainly interested by it. [I also] tend to – if any French or Italian film comes I will always go and see it cos I like to see what it's on. I like a whole variety of things – French and Italian films I will always go and see, I go a lot to the Watershed [Bristol's art house cinema].

(Abby)

The propensity among the dominant to mention these activities united young and old and male and female, but, as confirmed on the national level by Bennett et al. (2009: 123), varied by composition of capital, with those richer in cultural capital being over-represented. This suggests that the tendency to visit the storehouses of consecrated culture is propelled by the 'cultivated disposition' or 'cultural need' theorized by Bourdieu et al. (1991b: 37–70), that is, the attainment of pleasure through the application of elaborate schemes of classification to cultural artefacts that place them *vis-à-vis* one another through a deciphering of the symbolic 'messages' embedded within them, that comes with cultural capital. This inference is strengthened by the interviewees' accounts of their interest in one sphere of culture in particular – the visual arts:

I love going to art galleries, yeah I do, and sometimes if I'm in the mood I can sit there and watch a, you know look at a painting and really get to feel something of the art involved.

(Nigel)

I just absolutely adore knowing about buildings all around, knowing architectural styles, looking at something and being able to evaluate when and where it's come from and be interested in it, or going to an art gallery and *spending a bit of time thinking about the historical context of something and knowing what I'm looking at really*. I really, really enjoy it, yeah. That's my interest in art, I suppose. […] [my favourite style] would be pre-Renaissance Italian art. That was the last thing I went to see, in Venice. I find it fascinating, kind of religious art and how you can see a change from medieval art moving towards kind of full-blown Renaissance but not quite. It's very interesting to look at the styles and the altar pieces and how that then relates to the architecture of the period, so cathedrals and things like that and manuscripts. It's very interesting.

(Abby, emphasis added)

I prefer contemporary work, life drawing, Lucien Freud. I like oil paintings and multimedia always looks really effective. I think if you're going to have a painting, a work of art, *make it a work of art*, don't make it a reproduction of a photograph. If you want a photograph, get a photograph, that's my view. *I kind of like stuff that plays with materials more, plays with the oil or acrylic or whatever it is you're working with*. I quite enjoy Cezanne and stuff from that group.

(Debbie, emphasis added)

I quite like abstract art and some of the Impressionists' work. Modigliani I like, not particularly Old Masters, I don't like landscapes cos they don't do much for me. Probably more modern, contemporary stuff.

(Helen)

Sprinkled with confident references to artists and movements, guided by a proclivity for playfulness, embedded symbols of meaning and position and art that, if it is to be worthy of the name, issues from an individual 'creator' and is thus anti-realist and anti-representationalist, and indicating the requirement or propensity for leisurely and knowledgeable – in some cases even detached and, in the Kantian sense of disregarding all factors extraneous to the artwork, *disinterested* – contemplation, these exemplary accounts disclose in all its multiple manifestations the aesthetic orientation produced by the combined force of distance from necessity and cultural capital and their power to define what is considered 'leisure'.

The origins of this in the privileged school days of the 'confident amateurs', as Bennett et al. (2009) would call them, have already been traced, but it is also interesting to note that these interviewees are reproducing in adulthood activities that were revealed to be a matter of course in their childhoods, when their cultural capital stocks were being built, indicating the compounding force of familiarity with the institutions and practices of legitimate culture that comes from lifeworld experience. Diane, for example, recalls how she and her parents 'did all the museums' in her youth, including trips to London, and used to 'go to the theatre, go to Stratford, see Shakespeare plays', while Nancy declares that

> We always lived only half an hour on the train [from London], so I'd always been going up to the National Theatre and those sorts of things from a young age, fifteen or sixteen. My grandfather came to stay and we always used to go up to the National to see something.

'Regular and prolonged' consumption of consecrated cultural products is thus, it would seem, the product of an early exposure to the consecrated that is itself regular and prolonged (Bourdieu et al., 1991b: 109).

It should be stressed, however, that this disposition varies as position in social space descends towards the intermediate zone. Dominant interviewees with less inherited cultural capital, especially if upwardly mobile (for example, Lisa) or from mixed (for example, Sonia, Sue) or rural (for example, Rachael) backgrounds, expressed interest in theatre or foreign film, but enthused less over forms of culture commanding higher symbolic stature such as art, or even if they were interested their knowledge of and orientation towards valued cultural forms took a different form. Favoured artists or movements were more likely to be those, like Jack Vettriano in Sonia's case, occupying the pole of the artistic field concerned more with commercial sales and popular appeal than artistic integrity and critical esteem (Bourdieu, 1996b), sometimes based purely on a taste for visual beauty (for example, Rachael), at other times, as with Lisa's view on art, on a belief that:

> I don't think art should be about [the artist's self-obsession], I think it should be for the greater good, *or at least transmit meaningful messages*, you know. You know I think, sort of, Andy Warhol's interesting and made a lot of, I don't know sort of, he was quite a prophet really, quite prescient or whatever. But sort of contemporary art at the moment is like *I don't know what you're trying to say, you're saying nothing, which in itself isn't worthy* [laughter]. (emphasis added)

Acquired cultural capital here allows an articulate expression of the durable, lingering and fundamental *popular* disposition deriving from her dominated background for art that is 'instructive', harbouring moral commentary, or, in other words, fulfilling a *function*, even 'if only that of a sign' (Bourdieu et al., 1991b: 40; cf. Bourdieu et al., 1990: 86). In all these respects these interviewees – who would be described as 'dissonant' by others but who are in fact perfectly explicable with the logic of class once precise position, trajectory and lifeworld are taken into account – represent a step closer towards the dominated who, for their own part, display the polar opposite cultural dispositions to the dominant. None of them listed theatres, museums or opera houses as oft-frequented locations, and both Tracy and Gina, for example, expressed an arrant lack of interest in art and visiting galleries and, when pushed on their views on art, reduced it to a mild enjoyment of doing cross-stitch, paint-by-numbers and painting 'with children', that is, to the traces of 'art' in everyday, practical life. Others, however, were a little more open to the art world when pressed and had visited galleries, but found the art itself out of synch with their aesthetic sensibilities and thus displeasing. Chris, for instance, 'appreciates' landscapes and seascapes but cannot derive enjoyment from looking at pictures of 'fruit bowls' or 'people', suggesting a taste that is expressed more elaborately by Maureen and Doug:

> I personally don't think there's any skill in just getting a paint roller and flicking it around. [...] you just look at some things and it's like a four-year-old's done it. I like art but I just like it when there's skill involved, and I just don't feel that a lot of modern art has very much. [...] There is lots of different things out there, but from a personal

point of view, *I like proper paintings, where you can see what it is* – that is a horse, that is a dog. I just think there's more skill in that.

(Maureen, emphasis added)

I like pictures to look like a picture, you know, like your classics. Cubism I don't get at all, cos it's cubes. But if you look at a picture and it's a picture, like landscapes – Constable I thought was good, but I know he wasn't considered to be all that. Picasso, don't really dig that at all, like a big eye and stuff like that. If it's pleasing to the eye, I like it.

(Doug, emphasis added)

In opposition to the playfulness and impression or expression – in short, individual *creation* – valued by the dominant, the dominated aesthetic has at its heart realism, representationalism and artistic skill in *replicating* natural beauty, inclining them, it seems, to the 'classics' and Romanticism (without being named as such) and rendering most forms of modern art repellent. If the popular functionalist aesthetic manifested in a desire for instruction with Lisa, gratifying the 'interest of Reason', to use Kant's terms, then here it is manifest in taking pleasure from (or finding 'agreeable') the conventionally picturesque, gratifying the 'interest of the senses' (Bourdieu et al., 1990: 86; Bourdieu et al., 1991b: 40). Though they might, therefore, occasionally visit art galleries, perhaps as Bennett et al.'s (2009) 'relaxed consumers', they would not do so regularly enough – why would they, given that most of what is on offer is, in Doug's characteristically colourful language, 'shite'? – to count it among their named leisure pursuits.

Once again, however, members of the class are not all of a piece. Several of them – namely younger constituents with slightly higher levels of acquired (through post-compulsory art education) or mediated (through a lifeworld consociate) cultural capital – cited two preferred artists who appear to break with the usual taste for realism: Salvador Dali (and by extension surrealism) and M. C. Escher.[7] Gillian, for example, describes them as 'wicked', stating that 'every time you look at [one of their pictures] you see something different', while Josh (who has an A level in art) enthusiastically declares

I'm a fan of M. C. Escher and Dali, just crazy stuff. It makes you look at the picture and think about it. Rather than look at a landscape, where you look at it and go 'that's pretty', with this you have to look at it and you have to think. I think that's clever, brilliant, really good.

It is difficult to say for sure why these two artists in particular command such appeal, but two factors must be borne in mind. First of all, their frequent appearance on the school curriculum and, courtesy of educated 'cultural intermediaries' (cf. Bourdieu, 1984: 359; Featherstone, 1991), in popular culture (Dali's *The Persistence of Memory* in particular) may have bred the 'habituation and false familiarity' that Bourdieu et al. (1991b: 58) claimed popularize and thus devalue works. Secondly, both artists still produce recognizable *representations* – one of dream states, the other of impossible objects and scenes – but ones which, like the autostereograms that swept the shelves of bookshops in the 1990s, are treated as optical illusions *playing* with perceived reality in an entertaining ('crazy') way without (like Abstract Expressionism) negating it altogether. This may well be, then, little more than a variation of the fundamental class habitus, demonstrating that, even where formerly elitist art forms and works have been brought into the experiential worlds of the dominated, movements and works are, contrary to proclamations on the collapse of cultural boundaries and even Nick Prior's (2005: 135) theoreticist plea for a 'warped and accelerated' Bourdieu for new artistic times, still valued and appreciated according to the fundamental optic produced by precise position in social space, trajectory and lifeworld (including consociates and content of education). Even if the *precise* valued artists homologous with positions in social space have altered, therefore, they are still distributed according to the underlying relations of social difference and social distance that constitute class – a finding which, as we have seen, applies as much to appreciation of art as to tastes in sport, books, home-based leisure or whatever universe of practice one wishes to pick.

Class sense

So lifestyle practices are objectively patterned in accord with the fissures of capital. But are they *perceived* as such? Does the homology between social space and symbolic space, in other words, give rise to a widespread and pernicious class sense as symbolic goods and practices are read, decoded and judged by socially induced schemes of perception, or has it, as Beck and the others would claim, been drowned by a deluge of individualist discourse? Well, insofar as a sense of relational difference and similarity induced by the reading of practices and behaviours was pervasive and provoked a plethora of moral judgements and emotional responses, and insofar as even doubters of class sense had to surreptitiously summon perceived differences in order to question them, the solutions offered to these conundrums by the interviewees' accounts are hardly favourable to the individualization thesis.

Relational differentiation and the perceived homologies upon which it rests were signalled most frequently in the pejorative description of typified incumbents of positions encountered in the course of life. An exemplary instance is Lisa's recollection of her university days, probed with a Garfinkelian feigned ignorance of common prenotions:

Lisa: [...] for a polytechnic, there were an awful lot of really posh sort of, public school people who you know, I'm not saying just cos you went to public school you're necessarily posh, but that sort of – there were quite a lot of Sloanes, you know, who are really like [imitates voice] 'oh yah, marvellous darling', you know and driving around in sort of posh cars, and at the time had car phones and all this.

WA: What's a Sloane?

Lisa: Oh right, erm okay. It's someone who's – I'm trying to think of an equivalent – so I guess Prince William and Harry would be typical Sloanes. I mean I know they're royalty but they're very sort of, very upper crust, not just necessarily wealthy or public school education but very you know, into sort of horses and polo and – trying to think what else. You know that – does that make sense? Maybe they don't exist anymore, I don't know. But at the time it was a definite term, Sloane rangers, that kind of thing, and they would really sort of look down on anyone who wasn't posh – a very social set, I guess.

WA: Did you have much contact with them?

Lisa: No, not at all. I think I was quite antagonistic towards them really.

WA: Were they the majority or the minority?

Lisa: In reality at [the poly] they were probably the minority I guess, but they felt like the majority cos they tended to be quite loud and go round in big groups and make their presence known I guess.

In a comparable fashion, Zack makes reference to the prevalence of 'rahs' when describing the affluent area of Bristol in which he used to dwell:

WA: What do you mean by 'rah'?

Zack: Oh right, just erm, yeah it's Sloanes you know? Basically yeah, posh kids, rich kids, anyone who's got that intrinsic

WA: arrogance, I think is probably the sort of general description. You can usually spot them, which is quite funny.

WA: How?

Zack: There's a tendency for rugby shirts among men, that'll be pink. There'll be pashminas and things you know. Basically mini Paris Hiltons is what you can imagine for the girls, you know. It's a horrendous stereotype, but in a lot of ways it's true. My first year in halls I lived in one of the more expensive halls, so it did actually attract quite a lot of – well me and one other person were the only people on our floor to have not gone to private schools, you know, and kind of over time you built up a mental image that's quite difficult to explain, but it almost functions on the street where you go like [indicates pointing someone out in the street] 'there's one'. Yeah, I dunno how justified it is, but there you go. It's like racial profiling, but not.

Both accounts demonstrate the prepredicative, 'difficult to explain' pairing of affluence and private schooling (in other words, a certain neighbourhood of social space) with specific symbolic practices (polo, wearing rugby shirts) constructed from experience, both appeal to celebrity figures as model representatives of the classification in order to facilitate communication of the typification bundle (or 'mental image'), both mobilize extant linguistic descriptors produced and disseminated as part of the historico-symbolic struggle as epithets for the constructed category (rah, Sloane, posh), and in both cases the construction prompts an affective-moral response insofar as they patently recall with resentment the perceived arrogance, brashness and condescension – in other words, the brazen practice of symbolic violence – trussed to the typified individuals. Both narratives are no doubt so pellucid and elaborate because they are delivered from the mouths of upwardly mobile interviewees and refer to the disjunctive, 'disembedding' experience granted by the entering of worlds – the university, the affluent suburb – populated by agents from sectors of social space distant from their origins, yet, when pushed, *just about all respondents* reported making – these are pervasive refrains – 'snap judgements' or 'instant perceptions' based on appearance and articulation:[8]

[...] to see a student in Clifton [an affluent area of Bristol] who's eighteen and then see a girl in Whitchurch [a poor area in Bristol] who's eighteen and see the complete difference between the two.

You know there's a world of difference you know [...] you know you can tell what one [inaudible] just by looking at them, just by the size of their gold earrings or the size of their pashmina, you know basically that's – that and their hairstyle, you know, have they got all their hair scraped back, have they got a lot of highlights, have they just got blonde hair. You know it's quite easy to tell the difference without anyone opening their mouth what class they're from, even to what sort of job they maybe might be doing, you know. And with blokes it's the same you know it's very much dress, you know you can tell, instantly tell what class someone's vaguely, roughly from by the way they dress sort of thing.

(Andy)

I think a lot of store is put by voice and accent and stuff, which is maybe getting less so now but I don't know, there's always like the posh kids in school, and I think there's almost as much stigma attached to that as there is to being the other end of the stick. So yeah I suppose that's it mostly. Perhaps again where you live, people are gonna make a judgement about that, but yeah [...] Yeah, erm, there are always seen to be kind of fashions which are seen in certain groups of people. I don't know, like the classic Bristol shellsuit, lots of gold, and baseball cap or whatever – you know you would never expect to see that on somebody who's sort of high class. [...] And if they're well-spoken you're gonna think you know, they're upper class or whatever. Having designer labels and stuff with [inaudible], depends what labels you choose and whether they're genuine I guess. But yeah I think, I think everybody does judge from these sort of basics, it's not necessarily a good thing, but I don't think it's gonna change.

(Isabelle)

I don't know, I look at the kids I teach and there's a certain – you look at Vicky Pollard on 'Little Britain', I think that that idea hasn't come from nowhere, it's the kind of tracksuits with the stripes down here [on the arms], and the Argos jewellery, the big medallion style rings and ultra-necklaces and kind of piercings with gold jewellery. I suppose that is what you would look at, you'd be walking through Broadmead [the central shopping area of Bristol] and think – you wouldn't think that person was middle class would you I suppose, honestly speaking?

(Abby)

WA: Do you think you can tell someone's class from looking at them or talking to them?

Tina: Mmhm. Usually by their clothes, and if they've got great big Paris Hilton sunglasses on. Those things are so cheap. Most of 'em you can tell are all dressed prim and proper, average people just tend to dress average.

WA: What do you mean by 'prim and proper'?

Tina: Well, prim and proper, you've usually got the little skirts and their little things there and their little shoes and everything's all perfect and pretty. And you're like average like me, you've got a pair of jeans on, pair of boots and a t-shirt. [...] I don't actually like it cos to me it's a completely different sort of image. It's like, I don't know, too fancy-pancy. I get too dirty to sort of dress like that. Most expensive thing I've ever bought is my coat, and that's fifty quid. God, it was expensive! Burnt a hole in my pocket that did.

[...] people spend fortunes on clothes and they still look like trailer trash or whatever you wanna call it, chav style [...] people from poor backgrounds and deprived backgrounds, there is this stereotype of people who dress up wearing labelled clothing and all the rest of it because it looks like they've got the money to buy that sort of clothing. But they're a big joke really.

(Sean)

Whether dominant or dominated, whether picking out signature tastes, clothes or linguistic dispositions as markers of the typical habitus attached to certain sections of social space, whether distinguishing oneself from those perceived to be above (the 'fancy pancy' or 'posh kids') or below ('trailer trash' or 'chavs'), and whether addressing greater or lesser distances in social space (in some cases representing the small, 'last difference' that, as Bourdieu [1990a: 137] pointed out, can make all the difference in establishing self-worth) – all these examples, crammed with salient themes and by no means exhaustive, demonstrate in stark fashion the persistence and pervasiveness of relational differentiation based on the homology of social and symbolic space and, with that, cast doubt over the claim of individualization that class lifestyles and habitualizations, and the self-identification or self-understanding that they produce, have vanished.

Class sense is not, of course, always clear-cut, readily embraced or induced by the perception of apparel, but even where fuzziness and doubt surface symbolic differences still inform schemes of vision. Nearly all the respondents, for example, were keen to point out that their categories of perception were *fallible*, recounting actual or imagined encounters where they have misidentified people both above and below themselves on the basis of appearance and their typification bundles were proven 'wrong', 'unreliable' or 'over-generalizing':

> Taking a walk along the high street, you do stereotype people but – there may be a mum pushing a pram in a nice tracksuit, you know you put them into a certain class but talk to them and they might be 'oh hello, I'm Lady Asquith' or whatever. Unlikely, but you can't stereotype [...]
>
> (Mark)

> I see you wearing a shirt and tie, but for all I know you could be common as muck or something like that.
>
> (Josh)

> [...] when I was working one summer in a factory, I remember coming across a lad and I needed to ask him something. When I spoke to him he was quite well spoken, quite an articulate lad, and for some reason I was surprised and I thought, 'well why am I surprised by that?' I remember thinking to myself afterwards, 'that's really bad, you shouldn't have been surprised by that'. But I suppose because I was in a factory, I just assumed he was one of the usual, one of the people who worked there as a full-time job, and he probably wasn't. Might have been, but he may have been working there temporarily.
>
> (Liam)

Yet, in revealing how expectations can be confounded, these exemplary tales *presuppose the fact that expectations of behaviour and social standing based on specific symbolic markers have been built* and, until a real event notable for its discordance or an imagined event remarkable for its fictional status, have served as efficient ways of organizing the stream of experience. Indeed, so ingrained are the perceptual schemes that, in most cases, the jarring experiences are swiftly cast aside as exceptional ('unlikely', 'he probably wasn't') and their efficacy revalidated. Furthermore, some interviewees highlighted the obfuscatory role of strategies of self-presentation or 'semantic jamming', as Bourdieu (1987: 11) called it, but furtively implied that such strategies were not only limited deviations from established patterns but founded on a practical mastery of those patterns. Jackie, for instance, commented that class has nothing to do with clothing or image because 'people project the image they wish to', yet later demonstrated the principle of homophily by distancing herself from those who socialize in public houses (perceived to be below her) and those who are members of the local Lawn Tennis Association and collect classic cars (who are perceived to be above her) – in other words, those whose practices are removed in the symbolic space from her own pastimes and interests. Similarly, Joe claims that

> You can change your demeanour and the way you communicate and the way you act to the people you're around. I do it all the time, you change your behaviour to your environment. [...] People can wear what they want, can't they? I mean sometimes the way I dress people look at me and go 'you batty boy', cos I look gay. And I'm like 'what?'. I think clothing – you can change your look, you can change your clothes. I've got gear in the wardrobe and I'd dress up and people would think I'm a biker, and I can change my appearance and people think I'm a chav. Clothes are as changeable as the weather.

In almost parallel fashion, Caroline says:

> Caroline: I do think before it was quite stereotypical, you could look at someone and say 'they come from this background' from the way they talk, they act, they dress. Nowadays, you have someone comes from a posh part of London who dresses quite hippyish, so I don't think you can tell nowadays. You could dress really smart in a suit, learn how to put on a posh accent, and nobody could tell the difference that you came from somewhere else, like lower, you know what I mean. [...but] It's quite funny, cos we've got quite a few chavs who come to youth group, that I would class as chavs, with the jewellery covered everywhere, jogging bottoms and you think 'oh my days'.
>
> WA: So what do you think the usual background of a chav might be?
>
> Caroline: So stereotypical. It's not always, I wanna say this, but the ones that I know are from backgrounds where there's issues at home, with the parents. They do have a little bit of money, but not that much, and there's always some bigger issue that makes them want to be like that.

Both accounts, in an attempt to question the validity of class sense, present as evidence the fact that, with effort and strategizing, signs can be requisitioned and manipulated, to greater and lesser degrees of success, in a bid to transmit a certain impression. Yet the *very conditions of possibility of such strategies* are the existence of objective and sensed homologies between social and symbolic space, constructed groups and, ultimately, a hierarchy of symbolic value. The *modus operandi* of symbolic strategists, prompted by specific circumstances or social conditions (such as 'keeping up appearances', trying to 'fit in'[9] or even confidence trickery), is only enabled and informed by the assumption that certain signs and symbols (accents, clothes, jewellery), including those united in 'ready-made' categorizations sanctified as extant by both Joe and Caroline ('chavs'), are typically associated with certain social positions in the minds of the populace (having 'not that much' money). They simply would not and could not exist otherwise.

The reality of the matter, however, is that fallibility and fuzziness reared their heads only when the interviewees were responding to direct questions demanding a reflection on their schemes of perception alien to everyday practice. During the course of the life narratives, when the interviews approximated the conversations of quotidian life and summoned the practical (descriptive, explanatory) implementation of typification schemes, they were nowhere to be seen. On the contrary, though it was tied as much to broader differences of behaviour, attitude and orientation as consumption or leisure patterns, class sense frequently coagulated into a perception of sharp dividing lines and a feeling of existing in separate 'worlds' or 'realities' from others removed in social space contextualizing moments of life characterized by discomfort, disjuncture and crushed self-esteem. Lisa, for example, describes how the lack of money and material possessions *vis-à-vis* more affluent contemporaries when she was young diminished her sense of self-worth:

> I suppose you just felt not as good as other people really. Yeah, that you weren't as good as them and that somehow you were a bit different and a bit, as if you were from a slightly different world really, I suppose, not quite part of the modern world, if that makes sense.

The customary site for drawing clear cleavages and, feeling oneself on the wrong side, contused sense of self, though, is the *workplace*. Maureen, for instance, resented being 'looked down upon' by patrons when a barmaid, while Eddie recalls how, as a lifeguard, he was referred to as 'pond scum' by one of the managers and now, as a caretaker, feels a 'divide' between the maintenance staff and the teachers, the latter 'just walk[ing] past without a nod or acknowledgement' as if he does not exist and, again, 'look[ing] down on me'.

Two rather more elaborate accounts are provided by Caroline, talking about the 'snobby' affluent parents of the children she tends at work, and Tina, discussing the 'snotty' professionals in her workplace who subject her and other maintenance staff members to peremptory orders:

> I still see parents come in who are quite snobby, and I just think, 'oh dear, I'm running a mile', cos I don't wanna know people like that really [...] You *feel intimidated by some parents*, but you just think 'talk to them normally', and because I've shown they've not intimidated me, even though they have, you sort of gradually get to know them as a person and not so scary. You kind of can have a laugh with them a little bit, but *you're still quite wary of what you say* [...] You do tend to feel like *you're picking your words more carefully and trying not to sound common*, where with other parents you're completely yourself. [...] but you tend to say if someone is of a higher class to you, you would speak to them with a bit more respect [...] you have to communicate differently, use different, I would say jargon, with them, cos otherwise *they will think you're common and thick and stupid*. I think generally, not wanting to stereotype, but if you've got a really Bristolian accent, you tend to be treated like you're thick, even though they know you're not cos you've got your qualifications. It's kind if, *you have to try to get their respect*.
>
> (Caroline, all emphases added)

Tina: There's a strict line – we're scum, they're not. That's how it feels anyway. Not all of them are the same, I mean you do get some nice people. They're really sort of genuine, they make you a cup of tea, they sort of move out the way and hold doors. A lot of them they're just sort of, 'you're scum, I'm not looking at you'.

WA: How does that make you feel?

Tina: Shit. You get used to it, though. It's kind of funny after a while, you just – 'you're so pathetic'. But they wouldn't be anywhere without us anyway – if they didn't have anyone to keep the rooms that they work in, the buildings up together, decorate them, kept them looking good, then they wouldn't have a job. They forget that bit. Oh well.

In Caroline's case, intimidation, self-consciousness and self-correction, all in a quest to distance herself from a pathologized construction of those holding modest cultural capital ('thick and stupid') and to win approval from the (mis)perceived judges of, and thus superiors in, worth – even after sustained interaction and familiarity with the concrete individuals – indicate that this is a life afflicted by the insidious and incessant threat and exercise of symbolic violence. In Tina's case, a few pithy words express the invidious 'class contempt' (Reay, 1998a) on behalf of the privileged professionals, that is, the (mis)translation of class differences, via socially produced and situated schemes of interpretation, into perceived differences of *value* prompting behaviours, such as the 'tendency not to see or hear others as people' mentioned by Andrew Sayer (2005: 163), which make their targets feel 'shit' – a terse term that succinctly encompasses a vast montage of negative sentiments and self-evaluations – and, because of their powerlessness, 'get used to it'.

This is not necessarily a one-way street. Resistance to symbolic domination means class contempt is 'felt up as well as down', as Sayer (2005: 163) rightly notes and Karen's narration of her schooling, in the context of trying to understand why she worked so hard, demonstrates:

Karen: I didn't get on that well at school with some of the other kids. I suppose I didn't always have that many mates and you know, there weren't actually that many kids of professional, liberal parents in my school, if you know, like very, very few really. I think [we were] probably the only *Guardian* [liberal broadsheet newspaper] readers in a fifteen-mile radius [laughter]. It felt like that anyway. So I think it sometimes was a bit of a defence really.

WA: What was it that made you not get on with them?

Karen: I don't know, I suppose we just didn't – I did get on with them, I suppose I just didn't fit in really, always. I did have friends at different times, but I think I was quite different really.

WA: So what did the parents do of the kids that you knew?

Karen: Erm, I don't know really. A lot of, you know there weren't many kids of teachers around, there weren't many kids of you know, social workers, lawyers, that sort of thing, doctors. I think they probably went to private schools. And it just, it just felt like I was different, bit different. But at the same time I wasn't miserable all the time, and I did quite enjoy it as well. And I had some really good fun things that

I did. But it was just sometimes quite hard work, and I did
get bullied a bit sometimes. [...] I think like I said, having
Guardian-reader parents probably disadvantaged me a bit at
school because everyone else came from quite, most other
people came from such a different world, and it was quite
hard to understand where they were coming from for me.

Upwards class contempt, produced in this case by the disarticulation
between the dispositions generated in the familial domain of the lifeworld and those of the majority of Karen's school consociates, evidently
has serious ramifications for emotional experience and sense of self. Yet
it is tempered by the fact that, however scorned by her schoolmates –
who, no doubt, were trying to attain a fleeting sense of superiority
denied to them after their final school bell rings – it was Karen and not
them who possessed the capital and dispositions valued by the institutions she passed through and allowing her to succeed in the education
system and attain an economically and symbolically rewarded position
as a doctor, and thus she who, ultimately, emerged as the final victor
from the schoolyard struggles she has long since left behind.

Conclusion

The idea that the sun has set on symbolic differences between classes,
that lifestyles are reflexively chosen, detached from class positions, transient, fluid, eclectic, individualized, antecedentless *et hoc genus omne*, and
that class differences are no longer hierarchized, experienced, felt to the
nucleus of one's being and capable of arousing the most disturbing affective responses is as far removed from the social reality shaping the lives of
the interviewees as chalk is from cheese. From the privacy of the home to
the public institutions and activities of work and leisure patent class differences and their translation into experience and subjectivity steadfastly
endure. The aestheticization of household tasks versus the exercise of
practical mastery in the domestic sphere, the abstract versus the concrete
in reading, distinguished versus mass sports, taste for high culture and
abstract forms of art versus lack of interest and a taste for realism, and the
clear practical sense of and affective response to difference and distance
and inferiority and superiority – all these are so many expressions of the
fundamental division between a class habitus forged in milieus distant
from the pressures of necessity versus one compelled to live life according to the logic of first-things-first. Sure, there have been changes in the
concrete practices, goods and artists *representing* these tastes over time,

and of course there are complexities – the fine shades of class position, trajectory and lifeworld produce variations of the master patterns in art perception, for example, or the limited (and not necessarily new) omnivorous reading among the dominated – but nothing that breaks with the fundamental vision of relational class structures and homologies laid out by Bourdieu in *Distinction* thirty years ago.

7

'Class' as Discursive and Political Construct

So the distances and differences of social space and symbolic space continue to frame perception of practices and people and, therefore, shape tastes and pervade relations with others. Evidently class *is* still experienced and, accordingly, we can safely say that the reflexivity thesis has been refuted on that score. But some precision is necessary here: *theoretical classes* or classes on paper, the objective clusters in social space that map into symbolic space and shape perception, have been shown to mediate subjectivity, but the salience of *constructed* classes – the explicit discourse of 'class' as a means of grasping and articulating the differences of the spaces and fabricating social and political collectives that Beck and the others often have in mind when announcing the decline of class – has not been properly demonstrated hitherto. This is, in a sense, not strictly necessary: the core facets of class according to the Bourdieusian scheme have all been confirmed, and the fate of class discourse is essentially tangential. Yet if Beck and the others are to be assessed fairly – the absence of constructed classes might lend at least some credibility to their claims – and if we are to attain a grip on any contemporary trans-lifeworld doxa sustaining social division, this must be assessed. In this final empirical chapter, therefore, the task will be to examine the use of class labels as descriptors and typification bundles and – sure gauges of 'class consciousness' in the Marxist tradition, but here an indication of the importance of an established principle of perception – their linkage with political proclivities. Given the significant political shifts described in the opening chapter, namely the weakening of trade unions and the move to the right and jettisoning of class discourse, where it was present in the first place, among political parties of the left in most Western nations, the argument might at first appear more favourable to the reflexivity theorists than has been the case so far. It will soon become clear, however, that the processes unmasked, though liable to being misread by naïve sociological theory as testament to the absence of class, are in fact rooted in the differences, strategies and struggles generated by the structures of social space.

The uses of 'class'

The customary starting place for examinations of the salience of class discourse, particularly in survey research, is explicit self-identity, that is, whether individuals think they are in class X or class Y and how this corresponds with their occupation, resources and so on (see, for example, Surridge, 2007). In reality, the troublesome duo of direct questioning and fixed-choice answers this employs not only foists preconceived classifications (such as 'working class' or 'middle class') on people by delimiting options, thus forcing individuals to think that these categories are the legitimate way to divide up the social space in perception and inadvertently *perpetuating* their use when they may otherwise not be invoked, but prompts a reflective stance removed from the conduct of everyday life where the practical mobilization of relevant categories of perception in quotidian communication reveals their actual 'meaning' for users. A superior ingress into the latter is biographical narrative, where the prevalence and character of *unprompted* deployment of 'class' as a perceptual and linguistic construct in the concrete business of efficiently conveying experience can be investigated.

Following this route, we find that 'class' is, contrary to what the reflexivity theorists would have us believe, a popular discursive vehicle for accumulated typifications of the homologies of social and symbolic space, if an incredibly diverse and pliable one. All manner of behaviours, practices and politics were labelled 'working class', 'middle class' or, occasionally, more nuanced terms like 'lower-middle class' or 'upper-middle class', ranging from mild racism being the 'working-class way' (Doug) and Labour being for 'the working class' (for example, Jimmy) to gardening being a 'rather middle-class' pastime (Jackie) and complaining a 'middle-class' trait (Zack). This could even produce, from time to time, *interpellations* in the loose sense of recognizing oneself ('that means me') in the words and deeds of abstract others, as when Jackie bemoaned recent government health initiatives:

Jackie: [Politically] I think we need a bit of a change, and all this nannying about – what is it now? Middle-class drinkers, yes, apparently we're middle-class alcoholics. Have you read all this stuff?

WA: No, I haven't heard about that.

Jackie: Oh they've started on – because some people binge drink, they've decided to pick on middle England to say that we all drink too much. Which I'm sure we all do, but I think we're perfectly capable of making that decision ourselves, I'm not sure we need the government.

However, by far the most common use of class labels in biographical narrative, automatically entailing interpellation with designated categories and therefore clearly spelling trouble for Beck and the others, was to articulate one's 'place' in society and one's difference from others. This essentially took two forms, both of which were fairly widespread throughout the interviews. First of all, interviewees would use the terms as handy categorizations for making sense of people's – either their own or those of concrete or abstract others – *conjoined origins and dispositions*, usually rendered with the lay terms 'background' or 'upbringing', and thus for distinguishing practices, expectations, circumstances or trajectories that were similar to or differed from their own, especially at school, university or work. Helen, for example, claimed to have had 'a fairly average, middle-class up-bringing I suppose', in that 'everything was pretty ordered really' and finances were 'fine, very healthy' because her 'dad was earning an awful lot of money', which she then compared to contemporaries from the neighbouring housing estate: 'The children there would've been working class or unemployed, from unemployed families. Completely different set of experiences that they would've had.' Similarly, Barry states that 'I used to think of myself as sort of middle class, but more recently I've realized that if we were middle class we were lower-middle class in some ways, to put a label on it', while other children at his school were distinctly 'working class' with 'low incomes'.

From within the dominated camp, both Caroline and Andy render their sense of the achievable in education, of their 'place', in class terms, despite being a generation apart:

I imagine out of, I'd say out of everyone in the fifth year, I'd say about forty per cent of those people stayed on to sixth form and I'd say out of them maybe only ten per cent were people who eventually went on to university. People just didn't go to university. It would have been the more middle-class kids who went to university and they would tend to be the ones who stayed on in sixth form.

(Andy)

So I think that's another reason why I felt I would never go to university, I wasn't clever enough and never had enough money. I was working class and it was only upper-class people that went, or really intellectual people who got scholarships.

(Caroline)

These examples pertain to school, but of course the sense of difference, and the propensity to describe this in explicit class terms, persists long into adult life, whether that be university days and an alien professional life for the upwardly mobile, like Tessa, or the work-based symbolic violence recounted in the last chapter, as with Caroline once again:

I don't really have any, wouldn't really say I have any friends from working-class families, that I've met in Bristol. I mean it's not like they're really loaded, but you know, none of them are from sort of working-class backgrounds. Whereas at school, that was, most of my friends were. [...] [at the university I went to] I think, the numbers, the demographic is quite I'd say, there's a minority of working-class students. I mean of course I'm not in a working-class profession now, which is a bit strange, but I still do consider myself to be a working-class person. I don't know whether that's right or wrong, depends how you classify I suppose. But yeah, I think there's definitely a minority of working-class sort of people at [the university].

(Tessa)

I don't really want to use classes, but it is, you know. If somebody, like the snobs, come in, because they unintentionally are a higher class to you, you are going to talk to them differently. If you're like middle class and you've got someone lower class, you might unintentionally, in your head, think 'they're lower than me, I'm going to talk down to them a little bit more'. It's a two-way thing, if you think you're higher than somebody you will unintentionally talk down, so then it works both ways.

(Caroline)

Secondly, frequent use of the discourse of 'class' was made to articulate a sense of place based on *the areas people lived in* and *the schools they went to*. Here more than anywhere 'class' is a shorthand, quick-and-easy shared reference point for efficiently conveying the perceived homologies between social space, physical space and educational ethos that set the context for their lives, with whole rafts of people reduced to their typified common denominators or overriding tendencies for the sake of practical communication. 'Working class' could designate an area populated with manual workers (Paul), that was 'run down' (Debbie), where life was 'hard' (Doug) and blighted by 'low expectations' (Martin), while 'middle class' could mean an area filled with 'people with lots of money' but an 'unbalanced community' (Tessa) or a 'wealthy', 'very expensive' but not very 'culturally diverse' or 'culturally active' region which, nevertheless, oriented belonging and self-perception:

WA: So you say [your region and school were] 'white middle class'. Is that how you saw yourself?
Courtney: Yeah, definitely.

Occasionally interviewees would indulge in more fine-grained analysis, as when Nancy described her school as 'lower-middle class' in order to categorize and situate socially the fact that though the children were from 'reasonably comfortable backgrounds', some parents 'struggled' and there were fewer professional parents, largely because there were other 'posh' private schools nearby where children with 'serious money' went, which meant that the school consequently had 'low aspirations' and tended not to 'gear' people for university. Luckily, however, Nancy 'probably came from one of the most affluent families within the school' and, in wishing to become a barrister, was unusual in her expectations, placing her *above* her classmates in terms of means, aspirations and eventual trajectory. Perhaps Andy, familiar with the discourse of class from his political socialization, epitomizes the tendency for refined categorization best:

So I think a lot of you know, not all but quite a lot of [the people living in the area I grew up in] would have been sort of from that, you know form sort of working-class to lower-middle class, as opposed to middle-middle class or upper-middle class – which now is much more middle-middle class to upper-middle, but it was a lot more mixed.

All this, remember, is entirely unprompted, unambiguously indicating that class descriptors remain lodged within the everyday typification schemes of the populace as functional and relevant means of carving up and making sense of the social world in perception. True, the labels used are semantically variable for the interviewees as the typification bundles, first assimilated ready-made, are accented through idiosyncratic experience – what some would dub 'working class' others would doubtless label 'middle class' and vice versa – but the spontaneous deployment of class discourse is rife and shows no attachment to any particular demographic – the discourse of class is, historically, a nation-wide one – other than a slight over-representation among the upwardly mobile as they groped for concepts to make sense of their disjunctive experiences.

That said, around half of the interviewees *did not* use class labels until prompted, even if they plainly displayed a sense of difference and similarity clothed in a different vocabulary (such as, for example, Tina's sense of distance from the professionals at her workplace in the last chapter). This is not necessarily a novel situation – even when 'class' was supposedly more prevalent in national discourse its explicit terms were often absent among lay people and replaced by such expressions as 'our betters', 'the lower sort' or 'us and them' (as in Hoggart, 1957). In any case, once pressed,[1] *all* the interviewees readily recognized the discourse of social class and forwarded various perceptual categorizations, usually in a three-tier system ('working' or 'lower class', 'middle class' and 'upper class') but again sometimes distinguishing more finely graded sub-classes (such as 'upper-middle') for what Abby called 'in-betweeners'.

When asked to elaborate the *content* of these typification bundles – that is, what practices, behaviours, symbols and judgements they were paired and typified with and what the interviewees used them to denote – the answers varied significantly, but several key indicators cropped up time and again, either in vague or detailed descriptions, on their own or in a variety of combinations. These included, firstly, *occupations*, with the working class comprised of manual work (such as factory work or mining for Diane), 'menial work' in which people have to 'get their hands dirty and work hard for a living' (Doug), trades and 'lower down' jobs (as described by Maureen), the middle class constituted by 'professions' and landed aristocracy making up the upper class, though sometimes, among the dominant interviewees, there was a mastery of market research categories. Secondly, there was *income*, money, wealth or, in Courtney's words, 'general economic circumstances', and then, thirdly, *education*, with the working class being labelled 'less' or 'uneducated' while the middle class are 'educated', the latter being almost synonymous with possessing a degree. Fourthly, there was *lifestyle*, or 'general way of life' (Liam), whether identifiable through home ownership, size of house and garden, type of car, gym membership, newspapers read and clothes worn, 'politeness', 'manners' or 'grammar' (Josh) or simply in the relative terms of how good one's 'chips' are (Doug).

Finally, there were what are essentially *dispositions* but translate into lay phrases such as 'state of mind' (Doug), 'psychology' (Helen), 'personality' (Josh) or even the word 'class' itself, as in Wendy's description of *'nouveau riche'* individuals attaining money but lacking the 'class' to match it, all of which encompass differences in aspiration and perception but also issues around the continuity of the sense of 'self' between different circumstances:

> But I think it ain't always about what you do for a living, it's your state of mind, what you've been brought up in. I could go and get a job in an office and wing it in an office, but I'm still working class as a person.
>
> (Doug)

Helen: But finances aren't class, there's more to it than that. It's a psychological state as well as a financial state.

WA: What do you mean?

Helen: You can't suddenly – people would've been in a similar position to me in the upper classes and aristocracy, they would've suddenly lost everything, lots of people have done that. But they don't suddenly stop being an aristocrat because they've got no money. They're still upper classes, aren't they? But it's much broader now, I know we still have people from that level, but not so many of them. But the middle classes are much broader. You'd be exposed to different things and your psychology would change or your perceptions would change if you suddenly lost everything, from being an aristocrat to having nothing. You would possibly find it quite difficult to move in the same circles. I haven't found it difficult to move in the same circles. A lot of my friends when I was younger were still living in great big houses and I wasn't, and I wasn't suddenly excluded from that life but I didn't have it any more, which is quite interesting.

For Doug, who has weathered a steady reproductive stasis in social space (albeit with a brief stint in a moderately successful punk band), there is evidently a sense in which the classed orientations and perceptual schemes of his habitus are felt to be deeply sedimented, durable and transposable from one environment to the next, whereas Helen, invoking the changing fortunes in her youth as her parents' separation took its toll, is less certain, at first asserting the stability of dispositions before then alluding to the plasticity of the habitus in drastically altered circumstances.

All the forwarded indicators of 'class' bear witness to the continued use of the term to express discursively one's perceptual schemes, or, to be more precise, the package of typifications prepredicatively absorbed and refined *through* experience to *make sense of* experiences dictated by the objective homologies between the structure of capital, the occupational spaces and the system of lifestyles and dispositions. Though some individuals tended to emphasize one aspect (money, education or whatever) over another, there was no obvious correspondence between the attested yardsticks of constructed 'classes' and position in social space, contrary to Lockwood's (1966) much-criticized thesis. What *did* vary significantly according to position, however, were, firstly, which categories were described as 'normal' or 'average', such as when Zoe described the 'working class' as 'normal people' like her and her family who 'work hard for not much money' or when Courtney described herself as 'quite normal and average middle class' (cf. Savage, 2000, 2005); secondly, the degree of symbolic mastery of the discourse according to cultural capital (with several respondents admitting they had come across the academic concept of social class in college or university, thus perpetuating the theory effect); and, thirdly, the *affective response to* and *moral valuation of* the symbols or behaviours of different 'classes' as description soon, as if inevitably, shaded over into judgement and critical distancing (cf. Sayer, 2005).

By far the most energy and words were expended by the interviewees on the final dimension. On the one hand, symbolic violence – the denigration of the dispossessed – was frequently dispensed by the dominant, though its degree of subtlety and the precise practices targeted differed considerably. In Jackie's case it took the form of a visceral assessment of the geographical correlates of 'class' as residential areas were first picked out as sure indictors of the social and symbolic difference contained within the term and then measured against the expectations and standards of a capital-rich habitus focused on cleanliness and grandeur rather than, for example, community or diversity (cf. Southerton, 2002):

> But you know, you drive round bits of Bristol and they're like a bit yucky. If you drive round other bits and you think 'gosh this is nice, couldn't possibly afford to live here' so, you know, bits at the top of the hill in [the affluent area in which I live] you know we admire before we walk down the hill to our house. You know but we're not sort of living in Lawrence Hill or something, which I don't think is terribly nice.

For others, it emerges in the judgement of actions performed by the dominated *vis-à-vis* one's own aesthetic orientation and values (such as diligence, presentability and the love of learning), the former remorselessly shorn of their grounding in available capital and misperceived as some sort of reprehensible free choice, the latter consistently implied through vocabulary to be superior in every way:

> I mean in Ireland there really was an emphasis on education, that this was the way to get ahead, and here it's get on Big Brother. What the hell is going on? Or win the lotto, you know. Why not go to school instead? [...] I mean the tattoos and the earrings and all that sort of stuff, I mean that's a dead giveaway. Again I just look in horror, women who allow their children to deface their bodies and thereby limit their opportunities in later life. What for?
>
> (Nigel)

> I think that people should aspire to better things in their life like anyone, I think people should – cos there's so much wonderful cultural stuff out there you can do, education, open your mind to certain things and hopefully just *improve* your lifestyle and just make your life more *enriched*. (emphasis added)
>
> (Debbie)

From the receivers' point of view, on the other hand, the symbolic violence implicated in the discourse of class manifests in the regularly stated association of 'class' and its measuring rods with 'snobbery' and being 'looked down on', whether by strangers, family members or abstract others either far-off in social space or closer than they endeavour to insinuate. Serving as a discursive frame for the class sense unearthed in the last chapter, the heightened self-consciousness and blows to self-worth that typically result have already been charted. Yet, prompted to reflect on a practice usually simply lived through and reacted to, many of these interviewees now *refused to accept* the shameful badges of lower social worth pinned on them by others, following one or more of three defensive strategies: deflation, subversion and inversion.

First of all, purveyors of symbolic violence, especially those struggling to maintain differences over small social distances, were invariably deemed to be trying to be or thinking themselves to be 'something they're not', fighting to be 'up there' and getting 'above themselves' – a deflation of pretence or 'call to order' long present in the culture of the dominated, which both Hoggart (1957: 86) and Bourdieu (1984: 380) understood as an implicit appeal to solidarity and conformity.[2] Secondly, for some the deleterious effects of denigration were countered by an egalitarian desire which, bracketing social differences and leaving only humanity's universal physical constitution and functions, attempts to undermine the foundations of symbolic violence (cf. Savage, 2005: 939; Archer et al., 2007a: 228). No matter how 'brainy' one is (Tracy), how much money one has or how many 'exotic holidays' one takes (Jimmy), 'we're all the same', they would say, meaning that 'we're all black and white and blood and muscle and bone' (Tracy), 'even the queen has to go to the toilet' (Jimmy) and, to use Doug's characteristically lively language, 'their shit stinks the same as mine'. So 'why treat people differently?' asks Tracy, while Maureen concludes, 'I don't think anybody, because they've been born into the working class or whatever, should be looked down on just because they're not middle class.'[3] In one way this could be seen as a partial penetration, in Willis' (1977) sense of that term, of the arbitrariness of the distribution of symbolic value, though one induced by the need to defuse negative experiences and, indeed, claim the moral high ground. But the sentiment, being held out of necessity, is only weak – some of these interviewees were quick to pathologize those considered below them or to champion their own cultural traits – and, lacking the symbolic power to pose any substantial challenge to the hierarchy, remains an ought, not an is.

In this much it is similar to the third and final strategy in which, though rarely clothed in the language of 'working-class pride', many of the dominated attempted to fight against symbolic violence by inverting the symbolic hierarchy, that is, valorizing their own dispositions and discrediting those of the socially distant and different. Tina, for example, chided 'stuck-up snobs' who say, almost as a mantra, 'daddy, daddy, buy me this, daddy, daddy, buy me that', stating that

Tina: *I think the way we grew up, like not having loads, is done well.*
        I couldn't go, I couldn't say 'buy me this, buy me that', cos my mum used to say 'sod off! I can't afford it.' So *you have to work for your stuff,* you get things for treats, whereas you get some that say 'buy me this, buy me that' like my mate used to live round the corner. She said 'buy me this, buy me that', 'okay then, okay then, okay then', and she's just a spoilt little brat and she used to sulk when she didn't get her own way. I mean her mum took her out of school one day to take her shopping. There you go.
WA: So they don't learn. ...
Tina: *You don't respect nothing, do you? If you've got everything you respect nothing.* (emphasis added)

Others present analogous sentiments:

> I see some of the kids at [the private school I work at], because they're fee-paying, they don't care so much about their environment and about their belongings and stuff like that. They'll chuck stuff round, kick their bags around, leave it outside so anyone could pick it up and walk off with it. It seems to be a case of 'I've lost that, oh well mummy and daddy will buy me another one.' So I think people from the working class, middle class and up to a certain stage the upper class, *they do tend to respect themselves a little bit more and think 'I have worked hard for this, I am going to look after it.* If I wasn't something then I'll have to work hard before I get it.' Whereas those in the upper class think 'get mummy and daddy to buy me a new one, I don't really care about anybody else'.
>
> (Eddie, emphasis added)

> [...] you don't spend more than you need to, you're careful and you end up, to me *you end up having a much more real approach to life* rather than having people who are brought up in an environment where they kind of never really have to think about it and they always have nice clothes and can do what they want and can go skiing and whatever, and you never do and you feel inferior by it. But I think growing up in that is, *makes you more, just more ready I think for life in general*, because it's not like suddenly you know you really nested or you kind of brought up really cocooned and suddenly you're dropped in life with nothing else, it's probably quite a hard awakening. I don't know, it's maybe my judgement from being on that side of it. So I don't think it's ... *it's probably an advantage when you're a kid*, I think it's quite good for you just normal thing. And I suppose it doesn't even, for me doesn't make me pursue wanting money or living for that, you know it just makes me, *I appreciate what I've got*, just not working for money but working for what you like and, yeah just that really.
>
> (Bernadette, all emphases added)

Hard work, earning one's way, getting nothing for nothing, being 'real' and respecting oneself and one's accomplishments – these long-docu-mented hallmarks of the culture of the dominated (see, for example, Hoggart, 1957: chap. 3), adapted and reacting to material conditions of existence and, as Bernadette illustrates, often retained through social space travel, are elevated to the level of ethical supremacy and the practices of the dominant retranslated via this optic into an appalling, and thus subordinate, wastefulness and disrespect. But in this symbolic struggle, this tussle over the legitimate construction and ordering of the social world, the power is not with the dominated. Being not the possessors of symbolic capital, the captains of the field of power commanding media representation and political debate, they cannot successfully impose their fancied re-hierarchization as the accepted background doxa and so, as we already know, continue to disproportionately suffer deprecation in their most routine interactions.

None of this is to say there was no confusion, no doubt and no questioning of the concept of 'class' when the interviewees began to explore it. A sizeable minority, for instance, claimed that, with expanded education, the democratization of consumption and social mobility, exemplified by the proliferation of affluent celebrities hailing from the lower sections of social space (such as reality television contestants), the class system had become more 'fluid' (Lisa), 'fuzzy' (Emily) or 'less clear cut' (Sonia), that the term itself might be 'outdated', 'old-fashioned' or, to use the pro-Giddensian words of Rachael, 'traditional' and thus inapplicable today, and that it was now harder to distinguish classes according to once-conventional criteria like jobs, income or education. Furthermore, several of the respondents referred to a new, Baumanesque bifurcation between a benefits-dependent, 'sink-estate' dwelling 'underclass' (Andy) and a 'huge middle class' (Helen) of various degrees of affluence, the former sometimes described in a sympathetic way (for example, Helen) and sometimes in a scathing, resentful register reminiscent of Charles Murray himself (for example, Diane). Tellingly, however, these interviewees, as a glance at previous pages will reveal, were *just as prone as others to use 'class'* as an unprompted, practical descriptor of people and places and *still distinguished 'classes' on the basis of alternative markers*. The consequence was an abundance of contradiction and confusion:

> [...] the class system, for me I don't know if it exists. It certainly doesn't exist in the way it used to exist. The class system was invented and popularized at a time very different to today, so I don't know what makes someone a class, for example. What's working class and what's middle? I don't know what makes a middle-class person. But I'd say I was brought up middle class and I'm still middle class, in a middle-class profession.
>
> (Sean)

> [class is] an old-fashioned term that we're hanging on to because we haven't got another way of thinking about differences between us.

[...] It's a funny, sort of meaningless term really, but it's shorthand for saying something about your financial, educational, family background. That's my reading of it.

(Craig)

Rachael: I kind of feel like they [classes] don't really apply. I like to think that maybe they are old-fashioned, I dunno, I know they do still exist, the class divide.

WA: So you [earlier] said 'working class', 'middle class'. What do they mean to you, those labels?

Rachael: Well, I think middle class, I sometimes just use that to like, talk about my friends or talking in conversation about values and attitudes. Like valuing education, travelling, things like that. I guess to me it means values and attitudes.

Wavering between viewing class as anachronistic and recognizing its contemporary existence, between declaring its inapplicability and admitting it to still be a useful category of perception and discourse, and between questioning its survival and detailing the contemporary content of the constructs in their minds, these interviewees valiantly struggle to cohere two observations that appear contradictory and confused only because they deploy the substantialist logic of lay thought. On the one hand, when asked – in a way alien to ordinary interaction – to contemplate the macro-social issues bearing on typifications, they do (rightly) perceive shifts in the *concrete symbols* and *representations* of 'classes', but on the other, and jarring violently against this, they partially penetrate the persistence of *relations* of class *difference* and still employ the label in the *practical* business of comprehending and communicating the homologies that structure everyday experience.

The situation is similar for those interviewees who, when explicitly discussing and elaborating on the topic of class, took issue with its perceived lack of coherence – accusing it of being 'muddled' (Jackie), 'woolly' (Paul), a 'cipher' (Zack) or 'difficult to define' (Karen) – and, in extreme cases, pursued a doubt-inducing questioning of 'class':

Bernadette: But, but yeah it's a very, it's a very weird thing cos I don't even really know how to class people, I don't even know who's middle class, who's one ... it's really weird. I don't know, [a work colleague] was saying something quite interesting last time saying, like for example a builder who's got a lot of money, who makes as much money as a, like an IT manager, IT director something like that, they seem to have a different class but they don't – you know they have the same amount of money but they're in a different class range and [...] Yeah so that was, so yeah, yeah it's quite an important thing, it's just it's something quite hard to understand well, where the limits are and what it really means. But I think it's something quite important, I think it's to do as well with how you grow up and how you, just which, where you come from really – if you come from somewhere really like popular or more middle class or which school you went to, how you speak, express yourself, and it kind of, it shows ... and sometimes, so it's not really to do with money sometimes really to do with where you come from. I don't really know what is it behind all that, it's quite an important thing that how people relate to each other and how they, who they are. I don't think it's to do with money, I don't think it's just to do with where you're coming from or, if you come from a very bourgeois thing or not you know you can be, that state of mind I don't know. [...]

WA: So would you put yourself in class?

Bernadette: But, so what is it? How would you classify it? Would you really say there's like a ... would you classify it at all and how would you do it?

Dave: Now I, it depends what you mean by class you know, I don't have a rigorous definition of it myself but I think it's, you know, to talk about the working class as people do, I don't really know what that means. I don't know if I'm in it, if I've always been in it, if I've ever been out of it or what because it's a grossly over-simplified term to me, it doesn't actually tell me anything about somebody if you say they're working class. [...] I am what I am, I do what I do you know, lot of people would class me as working class because I'm a truck driver, and a lot of people would class me as middle class because I went to a grammar school, it just underlines my point I think – it doesn't tell you anything about anybody.

This is not an indubitable sign of the decline or irrelevance of class discourse to these individuals, some of whom had elsewhere used class appellations as descriptive devices. Instead it is an altogether expectable reminder that 'class' is, for agents, first and foremost a *practical* classification used to convey in a parsimonious manner aspects of perceived reality in daily life. As Bourdieu put it, echoing Schutz (1962: 93ff) on the incoherence of the practically oriented stock of knowledge:

> The representations which agents produce to meet the exigencies of their day-to-day existence, and particularly the names of groups and all the vocabulary available to name and think the social, owe their specific, strictly practical, logic to the fact that they are often polemical and invariably oriented by practical considerations. It follows that practical classifications are never totally coherent or logical in the sense of logic; they necessarily involve a degree of loose-fitting, in owing to the fact that they must remain 'practical' or convenient. Because an operation of classification depends on the practical function it fulfils, it can be based on different criteria, depending on the situation, and it can yield highly variable taxonomies.
>
> (Bourdieu, 1987: 10)

The uncertainty and misgiving thus stems more from the situation, in which the participants were, at this late stage of the interview, forced to become quasi-sociologists, where the incoherence of practical classification is translated into a stated incoherence of the concept *per se*. Almost scholastic reflection on mundane reality is not, as Phil demonstrates, a familiar or effortless task:

> Phil: But it's [class] quite difficult to define – I'm quite blunt as I said, so that is quite bluntly how I'd describe it, but I'm sure there's way, way more in-depth stuff than I could ever envisage to understand. There are things I know quite well, like my trade, there are things I don't very well at all, and that is social class. I wouldn't really have an understanding for it. I think it's one of those things that if you don't study it, like I don't, you see it but because you've seen it all your life you just accept it without really understanding it.

Classed, but not 'class', politics

So, contrary to what might have seemed the more plausible claims of the proponents of individualized reflexivity, the discourse of social classes remains a prominent, if hazy, scheme of typifications through which not only the social world but one's place within it, one's social identity, is thought, described and felt. The most fundamental of moral values, and the judgement of others that flows from it, are clothed in its vocabulary and struggles over *recognition*, to use Nancy Fraser's currently popular terms, waged in its name. But it is not readily regarded as a framework expressing material deprivation, iniquity or calls for *redistribution* – nowhere did the perception of 'class' involve an image of socio-economic injustice or struggle between workers and bosses. Could this be a sign of a more general perceptual stripping out of the material differences underlying the symbolic hierarchy in the face of individualization or the rise of reflexively considered 'life politics'? Would that mean that class is, as Beck and the others claim, no longer implicated in the struggles and stances over political matters? Only analysis of the final element of the symbolic dimension of class, contemporary political proclivities, will provide firm conclusions on this and, as we shall see, dispense with the notion of reflexivity and reassert the importance of class while allowing a glimmer of support for selected aspects of individualization.

There are three relevant lines of inquiry here. The first one is a direct engagement with the reflexivity theorists on their own terms – if they say that the importance of class for politics can be gauged in the relative prevalence of materialist rather than post-materialist issues, such as Giddens' 'life politics' and Beck's ecological politics, then we might as well put this to the test. The interviewees were, therefore, asked to describe the political issues they considered most important and which influence where their cross goes on the ballot paper. Unfortunately for the reflexivity theorists, the resounding reality – perhaps prompted by a sense of what was, at the time, an impending recession, but also given statistical foundations by Majima and Savage's (2007) quantitative research pre-dating this – was that nearly everyone readily reeled off not just materialist but baldly economic issues, often as an opening gripe but sometimes jumbled with a medley of concerns. Occasionally, and especially among the dominant with an extended symbolic mastery of social affairs, this appeared as frustrated concerns over inequalities abstractly conceived, including those played out through education and health; at other times, chiefly among the dominated, it was garbed in what might seem an individualized vocabulary – but really just one verbalizing the practical mastery of concrete experience – of 'what affects me', whether that be income tax, stamp duty, fuel prices or the cost of living. Karen and Chris represent each respectively:

> Social things, to do with, particularly to do with inequality, and I find that a constant frustration that our society's so unequal. And I see

it all the time with medicine, about how disadvantaged people are, and I just see it as, I just find it incredibly frustrating that our society is based on making money, and I would just rather not live with it, consumerist society.

(Karen)

Well, anything that's going to save me money, realistically. Taxing and that kind of thing is a big interest to me, and what's going to help with my family in the future. All of that is an interest to me, so I keep a keen eye on that. Those are my main concerns really, cos that's what affects me directly.

(Chris)

This is not to say that non-material issues did not figure at all – many also listed international affairs such as the Iraq conflict, human rights or the situation in Zimbabwe, and a substantial number voiced their interest in environmental issues, with the majority agreeing the latter were important, though not necessarily vote-determining, when directly pushed on them (with some discussing the practicalities of recycling while others mused on the global politics involved). In fact, a handful mentioned *exclusively* non-materialist, international topics, but generally only those, like Debbie, Barry, Rachael and Courtney, with enough capital to allow the space, freedom and mastery to consider affairs far removed from the business of living and surviving in concrete existence or else those, like Zoe, still to find independence from their parents and face the insistent pressures of necessity.

The level of materialism among interviewees indicates the extent to which manipulations of conditions of existence are explicitly yoked to 'the political' in perception, but, contrary to what Beck and the others imply, this is not actually the primary measure of the importance of objective differences in social space for politics. Instead, because political propensities are the product of an adaptation to, rather than a consciousness of, their conditions of existence, that mantle falls to the degree of *homology* between the space of class positions and the space of political position-takings, whatever specific issues feed into and provide the material for the latter. Accordingly, working on our terms now, the second line of inquiry must determine whether the old correspondence holds good or whether stances on any one issue, be they materialist of post-materialist, are 'dealigned' from class position and, as Giddens would have it, reflexively formulated given the abundance of information available.

So, if we survey the general patterning of proclivities, what fault lines, if any, emerge? Well, beginning with the dominant, there is, alongside the heightened interest in foreign affairs and global issues already mentioned, a clear split according to *capital composition*. On the one hand, those members of the fraction of the dominant class whose current capital stocks are composed of a greater weight of economic to cultural varieties display a right-wing, Conservative-supporting worldview in which the judgement is that 'we get taxed to death' (Nancy), there's too much 'nannying' (Jackie), too much suppression of 'free will and self-determination' (Samuel) and benefits so generous that they actively encourage free-riding indolence (Diane). In direct opposition, there is the (over-represented) Labour-backing, pro-welfare and equality-hungry liberal-leftism of those with more cultural than economic capital, expressed in disavowals of private education (Helen), consumerism (Karen) or capitalism (Lisa), endorsements of egalitarianism (Emily) and declarations that they 'don't mind paying a bit more tax' (Sean). The same general fissure reappears amid the dominated between, on the one hand, Labour supporters of various tints of red (from the adamantly anti-capitalist Andy to the pro-Blair Doug) and, on the other, Conservative authoritarian-traditionalists favouring severe (including capital) punishment in the name of law and order (Gary), stricter immigration laws (Eddie), reduced welfare benefits and low taxes (Caroline), with a few apolitical cynics who have not and do not see any reason to vote at all among their younger representatives. To give a fuller example of each:

> I know I've seen people struggle in our place and relying on the tax credits and the child benefit credits. Without Labour, you wouldn't have it, you know what I mean? I haven't needed it, but for people who need it, especially single-parent families that wanna work but need a bit more financial help, then it's there for them. So I see that as a good thing, cos it's helping people work instead of making them stay at home and rely on benefits.
>
> (Jimmy)

> Immigration, taxes on fuel and things like that. Those are the two things that get my back up most. When you see people coming in, I know they've had hard times and that, and they have trouble in their own countries and that. But when they come in and don't do a lot but get all the benefits and get shoved up the housing list when there's people worse off than them that suffer for it. I'm not racist or anything like that, but I do think they push their luck and get away

with a lot more than they should do. I think the government should clamp down on it, they're too soft with them as far as I'm concerned. They let them get away with too much, and if they didn't treat them so nicely they might think twice about coming here. I think there's too many human rights people, along those lines, saying they should all have equal rights and stuff. [...] There's just so many different ways they tax you, and whatever way you turn to try and save a bit of money for yourself, it just gets taken away. The working family tax credit, they say they won't tax you so much, but then they won't pay you as much with this. Makes it sound good, but when it comes down to it you're no better off anyway. It's a load of spiel really.

(Eddie)

Tina: I hate politics. Load of bull.
WA: You want to expand on that, tell me why you think that?
Tina: That is all I could say. All they do is lie out their arse to get where they wanna be and when they get where they wanna be, they just shit on you. That is all there is to it.
WA: Are there any issues in politics that you think...
Tina: I have not got the slightest idea about politics cos I've made no effort to know what's going on. I mean I don't really – it's one of those things you deeply don't care about. I say I couldn't care if they all died. I hate it.

Using Goldthorpe's unidimensional class scheme or the crude 'manual versus non-manual' measures so dear to psephologists, this would indeed appear as dealignment, with swathes of the dominant inclining and voting left and large numbers of the dominated veering right. From a gradational, multidimensional perspective, however, *this is precisely as expected*. A glance at Bourdieu's (1998a: 5; cf. 1984: 437–40) diagrammatic representation of voting tendencies reveals that the tendency to vote left or right follows a *curve* through social space, bisecting the dominated section of social space and bending up through the cultural fraction of the dominant. This accounts for two well-known facts in class theory. Firstly, the cultural fraction of the dominant class are less inclined to the individualistic acquisition and safekeeping of economic capital represented by the neo-liberal economics of their cousins across the vertical divide in social space owing to the fact that, thanks to their capital stocks, they generally value economic attainment less than cultural enrichment, disproportionately engage in professions demanding empathy – whether artistic or through the various social services making up the 'left hand of the state' (Bourdieu, 1998c) – rather than egoism and, because of an education slanted towards the arts and social sciences, possess a symbolic mastery of social affairs – that is, a perceived connection of some kind between individual fate and abstract social forces (cf. Parkin, 1968). Secondly, Conservative-voting dominated agents, usually dubbed 'working-class Tories' in prenotional parlance, are long-established figures and, far from being troubling to class theory, are well within the logic of adjustment to material conditions (Jessop, 1974; McLennan, 1985: 257). Usually more secure than their counterparts demanding parity of incomes and common ownership, mixing employment and self-employment and painstakingly saving capital, but insecure enough to experience the anxieties of cautious capital accumulation and the deviant behaviour of dispossessed social and geographical neighbours, the political habitus appeals for order, tradition, respectability and independence – issues central to the British right since Burke. As to the cynics in their ranks, Bourdieu long ago recognized the refusal of politics among those who, lacking the cultural capital to engage with it or feel their view is worthwhile or will be listened to, are refused from it (Bourdieu, 1984: chap. 8).

But position in social space does not work in isolation from one's sedimented past; trajectory, whether one of social stasis or movement, is always a compounding factor. This is manifest in the frequently forwarded declaration, seriously troubling any diagnosis of reflexivity, that party affiliation and political leaning cannot be explained as anything other than a product of an absorption and adaptation of the views that engorged the early lifeworld (Bourdieu, 1984: 439–40):

> But I'm a Labour man. That's cos me dad's Labour I suppose, and you sort of pick up off him. When you're watching news – I used to watch news from a very young age, I used to watch news with my dad before lots of kids did, cos you do it with your dad, you know. In my household it was like mum did the tea and dad sat down and watched the news like, you know. So I used to watch it with him, and he would be like – when it was the House of Commons he'd be like 'hear, hear, hear' when some Labour spokesman says something you know, and then it'd be 'bloody Tories' [laughter]. So it's gonna rub off, innit [laughter]?
>
> (Phil)

> I think my parents have always been interested in what's going on in the world, you know, international news, and they've always talked about it with us since then. Yeah I'm interested in it. And they're

quite liberal, so I guess they've passed that on to me and my brother and my sister actually.

(Rachael)

I think that's all part of a, very much from when I was growing up it was working-class people vote Labour and posh people vote Tory, and some people vote Lib Dem. And so I've always been very anti-Tory cos that's not what I'm about and you know, very much it's affected my, my family voting has very much affected my own, the way I go politically.

(Tessa)

A clear reminder that dispositions and schemes of perception are not produced in adaptation to material conditions of existence afresh from generation to generation, but involve the learning and reapplication of bodies of received practice and ideas sedimented over time and perpetuated by their delegates – politicians, unions and so on – to make sense of experience. This includes, as witnessed by Tessa's statement, instances where dispositions are carried forward through social space travel, and in some cases the absorption can be so complete as to baffle reflective thought:

Emily: So I think the family politics, apart from my father, has always been more left-wing and if I have voted I've always voted Labour. But I don't really know very much about it.

WA: Are there any particular issues that concern you?

Emily: Right and wrong really, I suppose. I just always feel that Labour is essentially right and the Tories are essentially wrong. But I really couldn't explain that in coherent terms. It's just a feeling that essentially left-wing politics are more about the whole and Conservative politics are more about 'me' and protecting the individual at all costs despite what [David] Cameron says.

Even with ample cultural capital, the political disposition has sunk below the level of clear articulation into practice without clear predication and so, in the dig for justification, to use Wittgenstein's (1958: §217) terms, bedrock is struck, her spade is turned and she can insinuate only that 'this is simply what I do'.

Political inheritance is by no means always, like a slick baton change between relay runners, direct and unquestioned. In some cases there is active resistance to parental politics, as when Helen claimed to have taken a radical turn in opposition to her deeply conservative father, and in others, such as Sean's, there is claimed to be research and consideration of political options. But with a little deeper excavation neither process vindicates reflexivity. Rebellion against elements of the early familial lifeworld, for example, certainly exist, as Sayer (2005: 30–5) points out, but, as Pierre and Emmanuel Bourdieu (ironically) analysed in *The Weight of the World* (1999: 507–13, 536–48), it has its roots somewhere – through reaction to suffering or pain, as Sayer would have it, but also, for example, through peers, discordant family relations or contradictory circumstances – and in Helen's case seems to be the product of the disjunctive experiences following her parents' divorce (attending state school, relocating and downsizing residence, living with a single mother), a strained relationship with her alcoholic father and radicalization at school through friends and pedagogy. Furthermore, though there may well be some element of conscious consideration of options, especially among those inhabiting the sparser zones between clusters in social space, this is no more than mundane consciousness. For as we have seen, and as is the case for Sean, current political leanings, and thus what is researched and consequently judged best, are the product of the intersection of the interest stemming from one's present position (such as ample cultural capital) and the weight provided by past lifeworld experiences (here an anti-Conservative bohemian mother and 'left-wing' father).

Only a crass reductionism would interpret this as evidence that class is the sole dictator of politics and all else, as Pultzer (1972: 102) put it long ago, 'embellishment and detail', for the picture is complicated further by a multitude of factors – for example, gender (with women more likely to vote left), religion (as in Caroline's case where it produced an anti-Labour attitude on account of their record 'on the Christian side of things'), field dynamics (for example, those further to the right in the intellectual field, with ties to industry and private consultancy, will be more likely to vote right) and lifeworld and biographical specificities (as already discussed in the case of Helen, above). Nevertheless, insofar as material issues remain important and position and trajectory in social space correspond to and generate *prises de positions* in political space, the objective divisions constituting classes on paper continue to suffuse politics. This only leaves one question left to answer: does the significance of class translate into political practice and the construction of 'class' as a principle of *mobilization*, that is, a rallying point with representative bodies, designated leaders and a discourse that converts perceived injustices into a collective project for change?

A full analysis of the activities of 'class making' and mobilization is beyond the tasks of this study, but their expression in the lives of the interviewees can be explored via two indices: the use of the discourse of 'social class' to frame political discussion and envisage a group-based political venture in the name of social justice, and, second, membership of and orientation towards those organizations so often claimed to be the representatives of 'the working class', trade unions. On the first of these, talk of 'class' did creep into political discussion, either unprompted or when explicitly questioned on it, but rather hastily and infrequently. Explicit reference among the dominated to 'working-class' collectivism, for example, such as Jimmy's gloss that he has 'always been a big Labour supporter, always will be' because they 'do more for the working class', was conspicuously rare. More common, by contrast, was the use of 'class' among the left-leaning dominant interviewees to articulate political concerns or changes, no doubt because they boast the symbolic mastery necessary to think the social in terms of an essentially academic system of classification and analysis, but sometimes less as a principle of social justice than social difference:

> I think it's very much of an issue, yeah definitely. Much more than, much more than anything else – well, not anything else but I think a lots of the issues from anything are much more to do with class issues than with lots of different things, lots of other things, you know like misunderstanding between people is often to do with that rather than anything else.
>
> (Bernadette)

Then again, explicit use of 'class' as a framework for making sense of political proclivities and activism is not a sufficient marker of the strength of constructed groups. In contrast to much of Continental Europe, the discourse has always been limited to a relatively small number of sometimes high-profile militant representatives (Arthur Scargill, for example) and intellectuals in the otherwise gradualist and amelioratory rather than revolutionary British trade union movement compared to more encompassing terms such as 'working people' (Gallie, 1978), and even this is likely to have been challenged by shifting membership demographics (see van Gyes et al., 2001; Bradley, 2002). So perhaps the second indicator, the extent to which interviewees – specifically dominated interviewees – are involved with trade unions, the corporate bodies perpetuating and framing struggle, and the extent to which they envision membership as a matter of pursuing *collective* goals, is the better one.

Low numbers mean findings must be provisional, but they are nonetheless informative. While current or erstwhile union members did occasionally make reference to 'workers needing protection' (Jimmy) or people 'sticking together' or being 'united' to achieve more than they would do as individuals (Maureen), the attitudes of members towards unions were at best *ambivalent, individualized* and *instrumental*. They were ambivalent because, even among loyal members, there were claims that 'you can look at both sides of the scale' because 'management needs protection too' and privatization and competition can be a good thing (Jimmy), that unions can 'help' *and* 'hinder' people (Chris), that they can be 'interfering' or a 'pain' and 'stick their nose in where it's not wanted and just make things more hassle than they need to be' (Eddie), that they are unnecessary 'if you have a good relationship with your employer' (Dave) or that they were once too powerful and the legislation against them in the 1980s was a good move (Phil).[4] Their attitudes were individualized and instrumentalist because, on the one hand, most turned to unions only to settle individual problems and disputes with particular members of management (such as Phil's 'nuisance' line manager) and, on the other, when strikes were called many regretfully shirked them because they 'can't afford to go without money' (Maureen).

However, as much as this demonstrates the weakness of interest in a collective project, *it is precisely the type of orientation to unions that was documented by the* Affluent Worker *team forty years ago* (Goldthorpe et al., 1968b: chap. 5; 1969: 166–70), meaning that, far from being specifically induced by fresh social, cultural and political dynamics, they represent little more than a continuation of an existing trend in post-war affluent capitalist society. That said, what *has* changed within the timescale of the reflexivity theories is the *extent* of union membership. Participation has dropped drastically in the UK since 1979, from over half the 'employee workforce' being signed up to well under a third (Gallie, 2000: 309; ONS, 2009: 60), and this is exhibited here in the fact that many of the interviewees actively spurned union membership, and those that remained faithful tended to be older and sometimes all-too-ready to describe the shortage of unionization among junior colleagues. Younger female respondents in particular – under-represented in the trade union movement for a long time, though with recent relative gains (Colgan and Ledwith, 2002) – eschewed membership as a 'waste of time' involving protracted yet futile wrangling:

I can't be dealing with that. It's just always political arguments all the time, it's just crap. Can't be dealing with any of that. Just go in, do your work, get paid, little circle, I'm happy with that. I wanna get paid for what I do.

(Tracy)

They sit there and chat and make decisions that never get followed up on, that seems a waste of time.

(Tina)

While a lack of experience of, and – owing to mundane jobs and investment in family rather than the male-dominated public sphere – importance attached to, work has to be taken into account in these instances, it would appear as if the same refusal of politics in general voiced above leads them to refuse what was historically constructed as their general 'voice'.

In sum, then, even if the use of class discourse to describe political projects and orientations towards unions show little sign of transformation, the fact that fewer people huddle under their protective wings – indeed, that they actively shun them – means fewer are or consider themselves to be represented by delegated leaders constituting them as a 'class' and speaking and struggling for them in that name. In this much, the mission of the 1979–97 Conservative government to enfeeble the trade unions and slowly remove 'class' from political discourse (witness Thatcher's 'class is a communist concept' epigram), left essentially unaltered by a reinvented Labour Party with its rhetoric of 'no more bosses versus workers' and retention of destructive trade union legislation, has had its effect, at least if these interviewees are anything to go by.[5] They still see materialist issues as crucial, and their views still generally correspond with their class positions, yet they uncouple this from 'class', and the latter, in turn, becomes an apolitical typification system referring primarily to symbolic violence, or a question of recognition, wrenched from its structural underpinnings. So here, at last, there seems a modicum of support for Beck and Bauman on individualization, though not reflexivity. Still, without any attachment to the notion of 'class consciousness' this matters little for demonstrating the importance of class for politics – indeed, the very symbolic struggle through which 'class' has been removed from the political register *is itself*, as imposition of a scheme of perception rooted in class position (in this case Thatcher's petit-bourgeois individualism) as the legitimate, doxic principle of division, *a class process* (see Cannadine, 1998). But it does mean that while classed politics are as strong as ever, 'class' politics, as a means of galvanizing sections of social space in the battle for collective social justice, has become rather scarcer in recent times.

Conclusion

Previous chapters have seen the outright refutation of reflexivity and its associated elements; at the risk of sounding repetitive, here the conclusion is much the same. It has been demonstrated that, far from being abandoned wholesale as a defunct classification with the withering of older symbols of class position, 'class' continues to be an oft-deployed and understood linguistic typification encompassing a range of phenomena which, because they frequently appear together in objective reality thanks to homologies (like wings and feathers, to use Bourdieu's example), are linked and associated in a more or less vague system of meaning. Not only that, but because agents occupy positions within the objective relations their viewpoint becomes a view from a point, meaning that 'class', rather than being ejected from reflexive self-perception or for that matter a tag for membership of some discretely bounded group, is, as Savage (2000) and others after him have also demonstrated, a tool for making sense of one's 'place' *vis-à-vis* others. Furthermore, because the differences of social space are, through the symbolic capital of the dominant, translated into rankings of worth and moral capacity, 'class' becomes a framework for expressing, describing and resisting the symbolic violence witnessed in the previous chapter.

How far this differs from the past is difficult to tell, but given that many of the themes surfacing above – the pairing of 'class' with a discourse of ordinariness, confusion and vagueness over the term, the appeal to common humanity and resistance to snobbery, for example – are also plainly evident among respondents in research of an earlier era, as reanalysed by Savage (2005, 2007), there seems to be a great deal of continuity. Doubtless the precise properties determining in perception whether someone is higher or lower in social space and graspable with one label or another have altered as social change has washed away old practices and patterns of living and ushered in the new, but the relational differentiation producing a sense of proximity or distance, similarity or difference, generates exactly the same strategies and struggles as it did forty, fifty and even sixty years earlier. This is not, therefore, the 'individual*ization* of class' that Savage talks about elsewhere – not just because 'class' was used as more than a peg upon which to hang an autobiographical story and to express similarity as well as difference, but because 'class' is used in essentially the same way as it ever was.

The same is true of political propensities: there is no evidence here that class positions and trajectories are anything other than fundamental shapers of outlooks and affiliations, that the materialist issues associated with class interest are no longer at the heart of concerns and voting tendencies or that the orientation to unions among members is notably different. The only phenomenon that seems to have altered is the strength of the perceptual–linguistic construction of 'class' as a political entity, that is, as an agglomeration of agents with the same interests, as fewer people associate with the corporate bodies likely to speak for and about 'class', albeit alongside new principles of division brought into the discursive fold with a changing workforce, such as ethnicity, gender, disability and so on: trade unions. The result would seem to be a depoliticization of 'class', as the reflexivity theorists might anticipate, but not a fundamental revision of its powers of relational differentiation and generation of practice.

8
Conclusion: Rigid Relations through Shifting Substance

'[T]he results of our enquiry', concluded Goldthorpe and his colleagues (1969: 157) in the closing pages of their final volume, 'are not at all what might have been expected had the thesis of *embourgeoisement* been a generally valid one.' Four decades later this succinct statement can be echoed, with some confidence, regarding the thesis of reflexivity. Absent is the hypothesized reflexive individual liberated from classed conditions of existence and dispositions, and little sign has been seen of the alleged decline of classed tastes, practices and discourse. On the contrary, whatever the age, no matter the occupational position, and whether witnessed in the tales of childhood, education, work histories, lifestyle practices, social identity or linguistic typifications, the firm grip of class on biographies and perceptual schemes has been shown to remain unbroken in contemporary Britain. Through theoretical scrutiny and empirical investigation, individualized reflexivity and its late modern counterpart have, therefore, been exposed as exaggerated and ungrounded accounts of human action in the current era. This does not, as has been repeated throughout the analysis, necessitate a denial of the broad mutations in economy and society addressed by Beck and the rest or, accordingly, an assertion that the consequences of class are identical to those of yesteryear. But even if we admit that elements of what would be described as the *substance* of class, its *manifestations* – that is, the actual symbols and practices attached to positions, whether educational pathways, occupational experiences or new lifestyle practices – have altered with the social context, the system of *relations* generating and differentiating them, and ultimately defining class, remains unchanged. The theories of reflexivity, on the other hand, being exemplars of what Bourdieu et al., (1991a: 20ff) called 'spontaneous sociology', that is, of sociological knowledge locked within the erroneous substantialist worldview and hence tantamount to erudite and elaborate prenotions, confuse the shifting signs for their enduring source and thus pronounce dead only what their epistemological short-sightedness prevents them from seeing.

Same formula, different figures

So the image of the classless, reflexive worker may be an illusion, but the motif of significant social change is certainly not. Whatever its problems in the minds of his critics, Bourdieu's (1984: 101, 171) comparison of class with an equation or formal model is thus apt in at least one sense: just like a mathematical procedure whose abstract formula stays the same whatever values are assumed by its variables and output, class continues to structure the *inputs* provided by the new context – namely, the struggles and balances of power in the economic, political and educational fields – and produce new observable *outcomes* as a result (cf. also Bourdieu and Passeron, 1990: end figure). The last four chapters have attempted to demonstrate this and, to pull the threads together, the various findings can be recapitulated here in summary form.

(1) The education system, the alleged central motor of both reflexivity and class reproduction, has undergone significant expansion and alteration in Britain within the last few decades. Neo-liberal government policies have sought to encourage post-sixteen and higher education in a post-industrial economy among growing numbers of young people, sanctioning new universities, funding initiatives and information campaigns that supposedly broaden the terrain of choice. Furthermore, with the reduction of traditional manual occupations allowing easy transition from classroom to shop floor via familial or social contacts, even vocational options are framed in terms of an active assessment of alternatives in search of some form of self-realization. However, statistics reveal that paths remain differentially distributed, and the present research goes some way to establishing why. Choices have not become equally reflexive for all, because *success at school* and, therefore, the *valuation* of school are not equal for all. Differences of parental capital – both economic, in paying for private education, and cultural, in providing the experiences conducive to educational achievement – continue to frame academic performance and subsequent orientations. For the dominant, the mastery of abstraction and symbolism inculcated from early years dovetails with a taste for it – a love of learning in

which the demands of the school system and self-realization are one – while for the dominated, the school becomes an institution of exclusion against which they develop oppositional attitudes and prize the practical mastery and bodily ability they do possess. Consequently, willing and perceiving themselves capable, the dominants' projected and actual trajectories are ones of straightforward transition through the academic route of A levels and higher education without conscious deliberation beyond the specific subject and university. Older means of achieving social stasis – direct inheritance of property and family businesses – may have been overshadowed by the inheritance and nurturing of cultural capital, but if anything the reproduction of privilege is thus *as stable as ever* as the upper regions of social space become characterized by what have been called 'normal biographies', that is, linear, unwavering and anticipated life courses at odds with the chronic volatility hypothesized by Beck. The dominated, on the other hand, yearn to escape education, experiencing it through their schemes of perception as an alienating, unpractical and expensive waste of time of less utility than pursuing their practical mastery to economic reward, even if the precise occupational destination is uncertain and subject to the reflections of mundane consciousness. In both cases, certain pathways in social space were simply barred from consideration as the subjective anticipation of likely futures attuned itself to the objective probabilities inscribed in their position.

Yet there are some originating in the lower sections of social space, where in previous generations university was a distant and unfathomable prospect, who have seized upon the expansion programme and ascended in social space. However, rather than prove the weakening of class constraints in late modernity or justify the meritocratic ideal of letting 'talent' succeed, the relational reality of class was present here too. On the one hand, the trajectories of the upwardly mobile were set in motion by experiential peculiarities and hidden advantages stemming from their parents' particular positioning in social space, demonstrating the power of relational differences to the last inch, but on the other, their upwards trajectories were hampered by the lack of capital *vis-à-vis* the more affluent counterparts they encountered in their social journey and were therefore disproportionately characterized by toil and struggle.

(2) Post-education life and, in particular, the sphere of work have also, like the education system, undergone considerable shifts since the mid-twentieth century. Changes of occupation, along with conscious

deliberation on mid-career options, were not uncommon among the interviewees, and, though it is hard to make any claims on a broad scale and there is a danger of caricaturing the past, it is possible that these could have at least been encouraged by recent economic policies demanding flexibility in a volatile environment. But even if there is more 'contextual discontinuity' than in the past – at least in sheer occupational terms – the image of attendant disembedding from class cultures and induced reflexivity that others have inferred from these conjoined facts is misleading. Movements within the topology of social space are not erratic, random or unpredictable. Where job shifts and deliberation of options occurred, they were the product of nothing more than mundane consciousness – a general attribute of human existence – contextualized by the capital advantages and accumulated classed experiences and skills furnished by their relative position. There is not the constant 'refashioning of self' and annulment of the past implied by the notion of reflexivity, including versions where reflexivity is stratified by class or, as for Sweetman (2003), a part of the habitus: in a capricious economic climate, some people are perhaps more open to the idea of changing occupation and are, on occasion, led to contemplate their options, but those changes and contemplations are limited by the possibilities inscribed in bodies and things and are, ultimately, far from habitual. They are guided by, but not part of, the habitus.

(3) This was even more the case with lifestyle practices. Older activities and symbols of position in social space had waned and, with technological developments, cultural shifts and the appropriation of distant ways of being, new products and practices had suffused lifeworlds, but, as indicated by quantitative research (Bennett et al., 2009), sharp class cleavages remained and, crucially, the principle of uptake accorded not with the model laid down by the reflexivity theorists but with that constructed by Bourdieu three decades ago. This was demonstrated through the examination of home-based activities, including reading, and the practices pulling people into the public sphere, whether sport or the arts. Moreover, the homology between positions and practices, but also between positions and other behaviours, moral outlooks and apparel, was perceived and articulated as a keen sense of social distance that, entwined with symbolic dominance, often turned into a recognized source of diminished self-worth. The intuition of relational difference, of living in a different 'world' from others, is the cornerstone of social identity and, being hinged on the topography of class, refutes Beck

and the others' supposedly more plausible claims that identities have lost their class character in late modernity.

(4) Of course the latter claim also applies to the specific use of class *labels* to describe self and others, and though this is only tangential to the Bourdieusian understanding of class it was investigated to fully exhaust the assessment of the reflexivity thesis. Contrary to what Beck and the others would have us believe, not only did people readily use 'class' to render the differences of social and symbolic space, but even those who did not forward class categorizations as typification bundles without prompting recognized the discourse of social class and, though there was semantic variability, hesitation and assertion of change, all laid bare the range of facets the labels covered for them, gauged the place of themselves and others in the social universe in those terms and used them as storehouses for moral judgements. Yet they did not, by and large, see them as politically significant. The late-twentieth-century national and international political mutations have operated to weaken the discourse of 'class' in the symbolic struggle to establish the legitimate principle of vision and division and mobilize sections of social space against certain injustices. In a theoretical model with no need for notions of 'class consciousness' this detracts little from the analytical power of class – this is demonstrated through the homology between positions and position-takings and the fact that the waning of class discourse from the political field is itself explicable in class terms – but it does signify the attenuation of one plank in the resistance against domination.

All in all, reflexivity as an explanation of the interviewees' lives past and present and the fate of class within them fell down on all fronts, even if some of the trends said to have produced it do constitute the new environment in which class continues to operate. In line with the epistemological vision upon which the study is founded, then, and when taken together with the multitude of studies that have and continue to demonstrate the workings of class in particular ways and in particular contexts and locations, we can safely conclude that the reflexivity thesis has been thoroughly confuted by all available evidence. Of course there is no shortage of alternative perspectives eager to declare class extinct or trivial, whether they be, ironically, the last vestiges of a moribund postmodernism or the currently voguish complexity theory, but given that Beck, Bauman, Giddens and Archer represented one of the more credible and influential challenges to the old notion, hopefully others will heed the evidence and begin to think through the *relationship between* class and whichever new processes they wish to emphasize.

In the meantime, while it must always remain open to challenge and change, Bourdieu-inspired class analysis need no longer preface its conceptual or methodological innovations with robust retorts to 'retreatists' and intentions to reinstate class – a trend that necessarily marked the early cornerstone texts (for example, Skeggs, 1997, 2004; Savage, 2000) but which has lessened as the stream of thought has gained momentum. Instead it can now devote more time to the exploration of hitherto underdeveloped directions (such as quantitative and historical analysis) but also something that has been absent so far: a sustained 'working at the foundations', as Husserl once said to Gurwitsch, that is, a securing of the programme of empirical analysis to a logically sound conceptual base by clearing up theoretical gaps (such as the relationship between class and work: see Atkinson, 2009) – an endeavour which, I would argue, requires a more thorough working with and through, rather than dismissal of, Bourdieu's concepts than has been the case lately.[1] Only with this will it effectively manage to topple the long hegemony of Goldthorpe's unsatisfactory position in the field of class analysis once and for all.

In defence of 'emancipation'

Finally, a word on politics. Some of the identified changes in the substance of class have been brought by advancements in communications and transport capabilities, some have been artefacts of the increasing interdependence of the global division of labour, but others have been the product of political visions and initiatives expressly designed to reduce inequality. It is significant, therefore, that the latter have failed to fulfil their stated purpose, but it is not difficult to see why. Take, for example, the shifts induced within the educational field by the Conservative government in the early 1990s. Though in good measure motivated by the desire to subordinate education, conceived as a factory of human capital, to the demands of the post-industrial economy – smuggled in proclamations that increased tertiary education would be 'key to national success' – the abolition of the 'binary system' polarizing polytechnics and universities to increase degree-awarding capacities and manipulations of the school system were also ostensibly (and no doubt sincerely) designed to 'widen opportunities' and ensure that all children were 'free to choose' – one might even add 'reflexively' – their post-sixteen options.[2] But, without denying that increased numbers of dominated individuals have attained degrees, in reality its primary consequence was a restratification of the field of higher education in much the same way as observed by Bourdieu and Passeron (1979) in 1960s France, with an extra trickle of dominated individuals pursuing mainly higher credentialization of existing practical, and therefore devalued, knowledge within devalued institutions. Embracing the meritocratic principle that '*natural* skills, talents, energy, thrift and inventiveness' must be 'released, not suppressed' (emphasis added), the Conservatives failed to address the *guiding principle* of the so-called 'free' choices and the *social source* of different skills and talents: the habitus, as expectations of the future and dispositions ultimately grounded in inequalities of capital. It is no surprise that class inequalities persisted, therefore, given that they were never actually addressed.

The situation has not improved more recently.[3] It could be argued that precisely what the Conservatives missed was taken up by New Labour in a bid to ensure educational success 'for the many not the few', to use their foundational slogan. Shortly before the global recession swept the political agenda, for instance, the then Prime Minister foregrounded social mobility and fingered the low aspirations of disadvantaged individuals as a prime hindrance, while the New Labour flagship Sure Start programme would appear to be aimed at equalizing cultural capital through such initiatives as, for example, information packs and leaflets for parents on supplementary home-learning techniques to improve their child's educational achievement. But, if looked at closely, the same shortcomings are present. The home-learning guidance of the Sure Start programme, for example, presupposes the parental ability, patience and time afforded by *existing* capital – both cultural *and* economic – beyond the levels afforded by the government's compensatory schemes (grants, credits and the like), as well as the kind of positive valuation of education (which logically assumes a devaluation of manual work), and therefore inclination to participate in children's education, witnessed only among the capital-rich. It therefore not only hopelessly endeavours to miraculously conjure cultural capital out of a vacuum, in contrast to the early immersion in a 'cultured' lifeworld enabling steady accumulation among the dominant, but, in a model exhibition of symbolic violence, it presses dominant orientations upon those with alternative experiences of education while 'responsibilizing' parents such that educational failure can be explained via tropes of 'bad parenting'.

The concurrent discourse on social mobility that framed such policies was no better. The key to raising aspirations, so the official rhetoric held, lay in instilling within children from disadvantaged backgrounds not just the 'work ethic' – the realization 'that there are jobs available if you make the effort' – but the 'learning ethic' – 'the idea that if you work hard and study at school there are great opportunities ahead and therefore you must take up learning'.[4] Again pushing the dominant values, the products of privileged conditions of existence, upon the dominated, this statement aimed to alter subjective aspirations *without changing the objective probabilities* furnished by in capital possession that generate realistic expectations of the future, with two likely consequences: a rejection of the 'learning ethic' by those rejected by it, subsequently interpreted as personal failure on the parents' and children's behalf, or a mismatch of expectations and likelihoods as children are exhorted to build lofty hopes, dreams and expectations that must, in reality, be demolished by the actual structure of the labour market (cf. Bourdieu, 1984: 143ff).

Labour's intention to foster the learning ethic among all children was, ultimately, contradicted by their commitment to the same flawed principle that undermined the Conservatives' policies: *meritocracy*, often masquerading as 'equality of opportunity'. Thus Gordon Brown spoke of creating 'a Britain where instead of talent wasted, effort unrewarded, enterprise stifled, potential unfulfilled, we see effort praised, ambition fulfilled, potential realised', 'a Britain where everyone, no matter what their background, should be able to rise as far as their talents can take them' and a Britain where social justice is 'expressed by social mobility, not compensating people for what they don't have, but helping people develop what they do have, their talents, their potential and their ability'. The social conditions of 'talent' and 'effort' were never addressed, and even if, unlike the Conservatives, Labour recognized that 'talents can take many forms and not just one – practical, creative, communication abilities, analytical intelligence as well', the fact that vastly contrasting social conditions produce 'practical' talents and 'analytical intelligence' was ignored. As long as this is the case, and considering these different 'talents' are differentially rewarded economically and symbolically, there cannot truly be 'equality of opportunity'. One is tempted, therefore, to suggest that only by lessening the disparities in distance from necessity, alongside support mechanisms extending those of the Sure Start programme, could a difference be made, with the additional effect that it might bite into the pernicious premise upon which meritocracy is founded: the existence of a hierarchy of worth, or *symbolic* capital, corresponding with differential remuneration as well as perceived 'talent'.[5]

None of this is to deny the significance of issues gathered under the label of 'life politics', but it is to suggest that, in opposition to Giddens' influential argument over the last fifteen years or so, the supposedly old-fashioned and secondary goals of 'emancipatory politics' retain their centrality in the sphere of political debate and that, against the Third Way, a more progressive perspective may hold the key to a society in which one's life is no longer furtively tracked and judged from birth. In fact, insofar as stances on post-material issues are themselves anchored in material and cultural conditions of existence, it could be argued that the inequalities addressed by emancipatory politics *take primacy over* life politics, for if truly democratic solutions to the latter, where all have the inclination, ability and information to participate effectively, are to be attained, then the former must surely be confronted. Unfortunately, in a time when 'class' has lost much of its political purchase and when the economic interests of those who command the field of power structure the field of politics more than the lessons of autonomous empirical research, the frustrating certainty is that the dominant political agents are unlikely to spontaneously take up this imperative challenge. That being the case, the formidable but vital task of the engaged social scientist, in concert with those who speak for and with the dominated – including trade unions (whether or not they use the language of 'class') but also kindred organizations, associations, movements and intellectuals working collectively within and across nations – is to persistently and vociferously resist, 'fire back' and make the scientifically informed case for social justice in the belief that, if not given by the laws of history, as Marx supposed, a more equitable society can nevertheless be achieved through reason, commitment and action.

Appendix: The Search Process

The ordered transition from hypotheses to empirical test in the text could yield the false impression that questions of method did not arise in the course of the research, were not considered or, if they did and were, that they are not worth mentioning. Nothing could be further from the truth, and so, in this Appendix, I want to address in turn three sets of methodological questions that, even if bracketed for the sake of presentation, must now be broached if the openness to scrutiny that makes social science worthy of the name is to be attained. First of all, how exactly can qualitative interviews, often regarded as unable to illuminate the operation of social structures, purport to uncover the relevant concepts in play, and what are the limits of their utility? Secondly, and importantly, what is the composition of the sample, how were the interviewees accessed and how did the interviews proceed? Finally, some may be curious as to the extent to which research based on fifty-five interviews with individuals based in one British city can claim to adjudicate the merit of theories held to have a national and international application.

Reconstructing lifeworlds, trajectories and habitus

The conceptual precepts laid down in Chapter 3 indicate that the objective of the qualitative interviews was to reconstruct individual lifeworlds, trajectories and habitus – the latter including dispositions and schemes of perception – and the nexus between them with a view to unmasking structural conditions. But how, precisely, was this task achieved, how does it relate to existing research traditions and what are some of its built-in limitations? To address all this effectively we can take each concept in the analytical triad in turn, beginning first of all with the endeavour of reconstructing lifeworlds.

Within the phenomenological tradition, from whence the concept originated, conducting research into lifeworlds essentially consists of harvesting vivid ideographic descriptions of the routine, everyday 'reality of common sense' (Berger and Luckmann, 1967: 34) as it is lived, experienced and interpreted by an individual, group, or wider collectivity. More specifically, it appears to follow Schutz in seeking to capture the 'natural attitude', conceived as the 'cognitive setting of the lifeworld', as it is 'embodied in the processes of subjective human experience' (Titchen and Hobson, 2005: 124) and, as a guiding principle, 'refrains from any causal or genetic hypotheses' (Berger and Luckmann, 1967: 34).[1] The approach adopted here, however, demands more. While description of the past and present quotidian milieu of the individual, as relayed in qualitative interviews, is indeed an essential starting point in order to understand the form of experience that has sedimented into the habitus (cf. Bufton, 2003), we must go *beyond* phenomenological description to uncover the objective social structures that have shaped and continue to shape that milieu – its objects, settings and consociates – and the experiences (and hence habitus) it yields without, of course, neglecting the formative impact of the particularities that necessitate the concept. As Bourdieu remarks, phenomenological analysis of the taken-for-granted 'is excellent as far as description is concerned but we must go beyond description and raise the issue of the *conditions of possibility* of this doxic experience' (Bourdieu and Wacquant, 1992: 73, emphasis added).[2] An understanding, therefore, of what was and is 'normal' or 'taken for granted' for agents (doxa), as indicative of what was and is perceived to be 'reasonable' and 'expectable', is crucial, but as they relate specifically to objects, practices, people, constraints and attitudes acting as manifestations of the different forms of capital possessed and their transmission or, conversely, of insecurity, choice and openness. In this way we can, *pace* the programmatic declaration of Berger and Luckmann, explore the genesis of the habitus and through it practices and trajectories.

Inevitably this is not a straightforward or unproblematic process. It must be borne in mind that a reconstruction of the lifeworld is precisely that – a reconstruction. Its key parameters and salient features – housing, schools, workplaces, parents, friends – and the resultant impact on the habitus can be gleaned, but the constraints of the interview situation and the limits and particularities of memory recapitulation mean that the picture of the lifeworld pieced together from the participant's account can only ever, like a map of a territory comprised of patches of greater and lesser detail, be a rudimentary and partial approximation to its rich complexity as a lived reality. But more importantly, perhaps, there is the thorny question of exactly how to move beyond the participant's description of their lifeworld – which, as linguistic typifications built into their habitus, are still analytically epicentral – to unveil the underlying objective structures that exist independently of their words and, indeed, have shaped their acquisition and use. Strict constructivists (including phenomenologists), after all, would claim that the agent's description is an active construction of 'reality' and that access to anything beyond that is illusory (see Roberts, 2002: 7). However, following the tenets of Bourdieu's constructivist realism, introduced briefly in Chapter 3, the descriptions must be treated as at one and the same time linguistic typifications conveying reality from the point of view of a subjective scheme of perception *and* in at least some sense pragmatic indicators of facets of the milieu demonstrative of the operations of relational structures that could be verified with alternative sources – documents, records, others' accounts – if needed.

The same applies to the analysis of biographies, and here Bourdieu (2000b) has explicitly warned against the so-called 'biographical illusion' that haunts much life-history analysis, that is, the failure to move beyond the subject's rationalizations and construction of events (their reasons or 'in-order-to motives', as Schutz would call them) to trace their trajectory through the space of objective relations and the cumulative effect this has on dispositions and attitudes (the 'because motives'). In order not to fall foul of such an illusion, therefore, the interviewee's account of the junctures and episodes from their neonatal years up to their present 'biographical situation' (Schutz, 1970b: chap. 7) are treated as a rough guide to the travels made through social space and homologous fields, bolstered by the accumulation of information on certain objective indices (occupation, pay, qualifications). In this way, the altering physiognomy of the lifeworld brought by the movements – horizontal, transverse, and even more complex combinations like curves – or the lack thereof and their resultant impact on the habitus can be considered along with the events that patently reveal the constraints or opportunities granted by a certain level and structure of capital.

Again, however, this is not an easy task. As with the investigation of lifeworlds, the reconstruction of biographical paths is limited by not only the peculiarities of memory recall – confabulation, exaggeration or just plain forgetting – which depend on the structure and content of the habitus *qua* stock of knowledge and the situational imperatives of the interview, but also the limited information that can be attained in an interview. A comprehensive understanding of the structure and history of each field the agent is and has been positioned in and of the fields in which the institutions they have passed through are and were positioned, for example, remains elusive. Nevertheless, interviews can, as Bourdieu argues, grant a general and 'genetic' comprehension of the participants by enabling an overall 'grasp of the social conditions of which they are a product', that is, a 'grasp of the circumstances of life and the social mechanisms' affecting them – including, in departure from Bourdieu's own strict approach, the salient formative experiences and events of the evolving lifeworld that fall outside the strict logic of fields – and the primary categories to which they belong and of the 'conditions, inseparably psychological and social, associated with a given position and trajectory in social space' (Bourdieu et al., 1999: 613).

This leads us to consider, finally, the habitus. In fact, much has been implied on the investigation of its genesis and characteristic dispositions so far, and indeed it is in the narration of life histories and the taken-for-granted world that the different facets of the habitus – expectations, orientations to the future, principles of action, tastes, schemes of perception and class sense – come to the fore. Here, then, only a few comments on some of the more direct ways in which it was tapped are necessary. First of all, the interviews elicited information on the lifestyle practices, tastes and political views of participants and their origins in the lifeworld with the intention of uncovering their roots in the experiences granted by the material and cultural conditions of life, therefore either confirming the homology between such symbolic markers and the social space or their reflexively chosen nature in the face of augmented choice and availability. Secondly, although the life-history component of the interviews endeavoured to uncover the practical typifications and associations mobilized by individuals, the interviews also explored in depth the participants' schemes of perception as they relate specifically to class – that is, what 'class' meant to them as a system of typifications, whether and how they perceived their life events, sense of difference and similarity and symbolic markers in class terms, and how important they saw it to be – and, again, their connection to positions, trajectories and experiences. The strategy here was, for the most part, to try and deploy a Garfinkel-inspired 'feigned ignorance' when faced with a typification bundle with common currency in order to break it open and explore its precise contents for the interviewees. However, in order to better capture whether class has any explicit salience for the participants without specific prompting and to avoid the kind of criticism levelled at Marshall et al. (1988) who did the opposite, this line of investigation obviously came *last*, after the narration of the life history and discussion of lifestyle practices and views, so as not to influence the terms used to relay them.

The sample and interviews

So if that is the 'how' questions addressed, then the next task is to detail *whose* lifeworlds, trajectories and habitus are reconstructed. The research sample consists of fifty-five individuals, the majority of whom, after insurmountable difficulties gaining access through employers firmly closed that door, answered a postal request for interviewees. This call for participants was conducted in two sweeps – one in the summer of 2007 and one in the summer of 2008 – and targeted three separate electoral wards of Bristol on the basis of their occupational constitutions, as revealed by statistics from the 2001 Census, in order to maximize the spread of respondents across social space (that is, one characterized by predominantly manual occupations, one predominantly professional–managerial, one mixed). In owing to a slight imbalance towards older, more affluent and educated respondents, the sample was then equilibrated through snowball sampling and a further small recruitment mailout to recent graduates of the University of Bristol. In all cases, interviewees were required to be in full-time work to take part in order to guard against an over-representation of the unemployed and the retired in the sample, who are more likely to have the time to participate and hence volunteer themselves.[3]

The final sample is shown in Table A.1. The interviewees are more or less evenly split in terms of gender,[4] and range from 18 to 53 years of age, thus capturing the experiences of a range of generations all said to be subject to the social forces inducing reflexivity. Unfortunately, though several interviewees were born overseas, only one, Humayun, is non-white. The sample is, therefore, unrepresentative of the UK population on this score and, regrettably, precludes systematic analysis of the intersection of class or reflexivity with race or ethnicity. In terms of social space, however, there is a wide spread. Armed with knowledge of the interviewees' origins and current capital (especially incomes, home ownership and value, education levels, cultural possessions, capital levels of parents and partners and number of dependants, in specific combination) and their place within the overall system of which they are a part (the local housing market but also average incomes, distributions of educational credentials and so on taken from the January–March 2008 sweep of *The Labour Force Survey*, an imperfect but sufficient source for the purpose), their positions and trajectories have been loosely categorized as follows. Those who

Table A.1 The research participants

Name*	Age**	Occupation	Indicators of economic capital***	Indicators of cultural capital****	Indicators of social capital	Overall current position/ trajectory in social space
Abby	28	Teacher (secondary)	c. £24,000–£25,000 p/a income 'after tax', home owner, no children	Degree, PGCE	Father: affluent businessman Mother: teacher Partner: teacher	Dominant/static
Adrian	40	Solicitor (partner)	£550,000 p/a income, home owner, part owner of legal firm	Degree, law conversion course	Father: dentist Mother: teacher Partner: lawyer	Dominant/static
Andy	43	Studio manager	£15,000 p/a income, rented accommodation	GCSEs, Youth Training Scheme	Father: office worker Mother: shop worker Partner: mature PhD student (cultural but little economic capital)	Dominated/static
Barry	47	Project manager/ lecturer	£30,000–40,000 p/a income, dividends from his company	PhD	Father: accounts worker Mother: teacher Partner: school secretary	Dominant/static
Bernadette	32	Graphic designer	Income not divulged, no children	Degree	Parents: poor farmers Partner: carpenter	Dominant/ upwardly mobile/ from France

Caroline	25	Nursery nurse	£12,000 p/a income, lives with mother, no children	Unfinished degree	Father: routine IT worker, deceased Mother: home carer Partner: youth worker	Dominated/static
Chris	33	Customer services adviser	£15,000 p/a income, home owner	BTec Diploma	Father: health and safety adviser Mother: self-employed cleaner Partner: Tracy Also has affluent uncles (an architect and a company director)	Dominated/static
Claire	38	Senior manager	Income not divulged, home owner	Degree	Father: self-employed ophthalmic technician/optician Mother: administrator Partner: financial adviser	Dominant/static
Courtney	32	Managing director	£30,000 p/a income, plus income from let property, home owner, no children	Degree	Father: civil engineer Mother: computer analyst No partner	Dominant/static
Craig	35	Clinical psychologist	£58,000 p/a income plus fees for private work, home owner, no children	PhD	Father: pharmaceutical researcher/entrepreneur Mother: optical receptionist Partner: senior researcher	Dominant/static

(Continued)

Table A.1 Continued

Name*	Age**	Occupation	Indicators of economic capital***	Indicators of cultural capital****	Indicators of social capital	Overall current position/ trajectory in social space
Dave	51	Lorry driver	Income not divulged, home owner	Degree	Father: routine white-collar Mother: nurse Partner: mature PhD student (cultural but little economic capital)	Dominated/static
Debbie	36	Writer/ broadcaster	£30,000–£40,000 p/a income, home owner, no children	Masters degree	Father: affluent entrepreneur Mother: housewife Partner: pilot	Dominant/static
Diane	40	Entrepreneur	Income not divulged, home owner, no children	A levels, secretarial qualification	Father: affluent harbourmaster Mother: secretary Partner: musician	Dominant/static
Doug	43	Self-employed plasterer	Variable, but up to £30,000 p/a income, home owner	Unfinished apprenticeship	Father: electrician Mother: housewife Partner: primary school teacher	Dominated/static
Eddie	37	School caretaker	£15,500, home owner	O levels, CSEs, City and Guilds	Father: mechanic/office worker Mother: housewife Partner: housewife, formerly a bank clerk	Dominated/static

Name	Age	Occupation	Income/housing	Qualifications	Family background	Classification
Elizabeth	39	Computer programmer	£30,000, home owner, no children	Degree	Father: wealthy farmer Mother: teacher Partner: 'builds hospitals'	Dominant/static
Emily	45	Personal assistant	£20,000 p/a income, plus income from jewellery making, home owner	Masters degree	Parents: affluent publicans Partner: teacher	Dominant/static
Frank	53	Hospital technician	£16,000 basic p/a income, home owner	Apprenticeship	Father: technician/FE teacher Mother: typist Partner: home carer	Dominated/static
Gary	44	Driving instructor	Income not divulged, home owner	Vocational qualifications	Father: engineer Mother: secretary Partner: figure skater	Dominated/static
Gillian	29	Teaching assistant	£12,000–£13,000 p/a income, home owner	GCSEs, currently doing an NVQ	Father: lorry driver Mother: office worker No partner	Dominated/static
Gina	30	Learning support assistant	£9,000 p/a income, council accommodation	Currently studying for a degree	Father: bus driver Mother: postal worker/shop manageress Partner: warehouseman	Dominated/cusp of upward mobility
Hannah	30	Administrator (part-time)	Income not divulged, home owner	'Business management and secretarial diplomas' at a local 'small technical college'	Father: salesman Mother: charity worker Partner: telecoms engineer/supervisor	Dominated/static

(Continued)

Table A.1 Continued

Name*	Age**	Occupation	Indicators of economic capital***	Indicators of cultural capital****	Indicators of social capital	Overall current position/ trajectory in social space
Helen	42	Business support manager	£25,000 p/a income, home owner	Degree, PGCE	Father: sales director Mother: housewife No partner	Dominant/static
Humayun	31	Councillor	Income not divulged, home owner	Degree, unfinished Masters	Father: army officer Mother: housewife Partner: graphic designer and social services	Dominant/static/ from Bangladesh
Isabelle	26	NHS scientist	£23,000–£24,000 p/a income, home owner, no children	PhD	Father: NHS scientist Mother: teacher Partner: website designer	Dominant/static
Jackie	38	Project manager	£40,000 p/a income, home owner	MMus, MBA	Father: engineer/ manager Mother: teacher Partner: IT consultant	Dominant/static
Jimmy	47	Postal worker	£32,000 p/a income including overtime and bonuses, home owner	Apprenticeship	Father: window cleaner Mother: shop worker Partner: teaching assistant	Dominated/static
Joe	35	Technician	Income not divulged	Apprenticeship	Father: van driver Mother: administrator No partner	Dominated/static

Name	Age	Occupation	Income/Living	Education	Background	Classification
Josh	25	Sports coach	£19,500 p/a income, lives with parents, no children	A levels, GNVQ	Father: postal worker Mother: nursery nurse Partner: nursery nurse	Dominated/static
Karen	28	Junior doctor	£27,000 p/a income, no children	Degree, graduate entry medicine	Parents: social workers Partner: museum curator	Dominant/static
Liam	25	Economic development officer	£21,000 p/a income, rented accommodation, no children	Degree	Parents: social services/FE lecturers Partner: civil servant	Dominant/static
Lisa	34	HR officer	£22,000 p/a income, home owner, no children	Degree, PGCE, PG Dip	Father: draughtsman Mother: factory worker Partner: web developer	Dominant/upwardly mobile
Mark	35	Computer programmer	£28,000 p/a income, home owner	Degree	Father: university professor Mother: teacher Partner: manageress	Dominant/static
Martin	34	Surgical nurse	£31,000–£32,000 p/a income, home owner, no children	NVQ, City and Guilds	Father: heavy plant driver Mother: housewife Partner: surgeon	Dominant/upwardly mobile
Maureen	43	Housing adviser	£21,000 p/a income, council accommodation	GCSEs, YTS	Father: civil engineer Mother: home carer Partner: mechanic	Dominated/static

(Continued)

Table A.1 Continued

Name*	Age**	Occupation	Indicators of economic capital***	Indicators of cultural capital****	Indicators of social capital	Overall current position/ trajectory in social space
Nancy	46	Barrister (senior junior)	£120,000–£220,000 p/a income, home owner, no children	Degree	Father: company director Mother: teacher Partner: property developer	Dominant/static
Nigel	45	University reader	'high 40s' p/a income, home owner	PhD	Father: doctor Mother: undisclosed Partner: journalist	Dominant/static/ from Ireland
Oliver	40	Operations manager	Income not divulged, home owner	Apprenticeship	Father: naval commander Mother: teacher Partner: Inland Revenue worker	Dominant/static
Paul	41	Software developer	£30,000 p/a income, home owner, no children	Vocational qualifications	Father: electrician Mother: secretary Partner: HR manager	Dominant/upwardly mobile
Peter	49	Engineer	£65,000–£75,000 p/a income, home owner	Apprenticeship	Father: teacher Mother: housewife Partner: home carer	Dominant/static
Phil	33	Shift co-ordinator	£28,500 p/a income, home owner	BTec, ONC, HNC	Father: electrical engineer, Mother: housewife No partner	Dominated/static

Name	Age	Occupation	Income/accommodation	Education	Background	Category
Rachael	26	Project coordinator	£22,000 p/a income, rented accommodation, no children	Masters degree	Parents: social workers and farm owners Partner: traffic engineer	Dominant/static
Rebecca	30	HR adviser	£30,000 p/a income, home owner, no children	Unfinished degree, PGDip	Father: Catholic dean Mother: teacher Partner: business consultant	Dominant/static/ from Canada
Samuel	35	Doctor (hospital surgeon)	£60,000–£65,000 p/a income, home owner, no children	Degree	Father: prison officer Mother: housewife Partner: doctor Also had an uncle who was affluent through marriage	Dominant/upwardly mobile
Sean	36	Software developer	£32,000 p/a income, home owner	Masters degree	Father: architect Mother: artist Partner: publishing manager	Dominant/static
Sonia	25	Social worker	£26,000 p/a income, rented accommodation, no children	Degree	Father: social worker Mother: carer (divorced when young) Partner: garden contractor	Dominant/mixed origins
Sue	42	Social worker	£25,500 p/a income, home owner	Degree	Father: affluent businessman Mother: social carer (divorced when young) No partner	Dominant/mixed origins

(Continued)

Table A.1 Continued

Name*	Age**	Occupation	Indicators of economic capital***	Indicators of cultural capital****	Indicators of social capital	Overall current position/ trajectory in social space
Tessa	28	Junior doctor	£37,000–£38,000 p/a income, rented accommodation, no children	Degree	Father: lorry driver Mother: on disability benefits Partner: financial services (temping)	Dominant/upwardly mobile
Tina	18	Apprentice painter	c. £13,000 p/a income, lives with parents, no children	Currently doing an apprentice-ship	Father: bus driver Mother: cleaner No partner	Dominated/static
Tracy	27	Transport clerk	£12,500 p/a income, home owner	GCSEs, NVQ	Father: manual worker Mother: carer Partner: Chris	Dominated/static
Trisha	37	Technician	Income not divulged, home owner, no children	Apprenticeship	Father: electrician Mother: cleaner No partner	Dominated/static
Wendy	48	Assistant head teacher	£48,500 p/a income, home owner, property owner, no children	Masters degree	Father: avionics electrician Mother: market researcher No partner	Dominant/upwardly mobile

Yvonne	42	Driving instructor	Variable, but usually less than £4,000 p/a income plus benefits and income from occasional dance classes, council accommodation	Secretarial qualification	Father: builder Mother: secretary No partner	Dominated/static
Zack	28	Software engineer	£31,000 p/a income, rented accommodation, no children	Masters degree	Father: engineer Mother: on disability benefits Stepfather: salesman Partner: financial services manager	Dominant/upwardly mobile
Zoe	20	Clerical worker	Income not divulged, lives with parents, no children	NVQ	Father: postal worker Mother: office worker/ cleaner No partner	Dominated/static

*All names are pseudonyms.

**All ages have been changed by one or two years either way.

***Where incomes have not been divulged the average income of the occupation, as given by the *Labour Force Survey*, has been taken into consideration. Not having children has been listed as a boost to economic capital resources, and where this is not mentioned it means the interviewees had children of various numbers and ages.

****GCSEs, O levels, NVQs, City and Guilds, YTS, BTecs, HNCs, apprenticeships and job-specific credentials represent basic or 'vocational' options, or technical capital, whereas A levels, degrees, PGCEs, Postgraduate Diplomas, Masters degrees and PhDs represent 'academic' qualifications.

have comparatively little in the way of economic, cultural or social capital in the round and hence inhabit varying positions in the lower to lower-middle section of social space are described as *dominated*, whereas others with much more capital as a whole represent a mix of more or less *dominant* positions ranging from the middle belt of social space to the higher reaches, and from those possessing more cultural capital and further to the left of social space to those with more economic capital and further to the right.[5] Furthermore, trajectories have been categorized as representing either stasis or upwards mobility – the only two exceptions being Sonia and Sue, who both clearly had mixed, or 'cross-class', origins and, though their capital-rich fathers were not absent, had greater contact with their poorer mothers after their parents divorced – and migration between national social spaces has been noted in those cases where it has occurred. In reality, of course, positions and trajectories are complex affairs with finely distinguished locations and turns – as the analysis endeavours in some cases to show – so the labels attributed are purely reductive heuristic devices aiming to isolate the critical division of resources. It is for this reason that, while an 'intermediate class' (which Bourdieu labelled petite bourgeoisie) may exist when social space is mapped fully (as in Bennett et al., 2009), for analytical purposes this has been sliced in two and each half attributed to its closest neighbour in the binary split.

The main indicators of capital and, via parental capital, trajectory are listed in Table A.1, though even these mask the multitude of differences and details, excluded for purposes of clarity, that set people apart in social space and have played their part in the task of categorization – for example, the size and location (and thus approximate value) of the home, parental and spousal educational levels, disciplines qualified in, type and field position of the university attended and grades achieved, all but the most pertinent of family members and so on. Note that the possession of a degree is not an automatic or necessary entry pass into the dominant on its own, nor is above-average income. Thus, on paper, Phil currently has a higher income than Diane, an entrepreneur 'struggling' through the early stages of her product launch, and possesses a qualification officially designated as higher than Diane's A levels, but because he owns only a 'little two-bed flat' in an area populated mainly by manual workers, pays child support for two children as well as the costs of looking after them when he sees them, in actuality possesses a Higher National Certificate furnishing *technical* capital (in plant engineering), which he gained during his apprenticeship and has no partner or capital to speak of emanating from his parents, whereas Diane owns a large house in the most expensive area of Bristol (itself one of the most expensive cities in the UK for housing), which she can rent part of out for profit, has no children, has very wealthy and generous parents and a musician partner and has had an exclusive private education, the resources available to him *on the whole* place him in a significantly lower position within social space. Similarly, though Dave has recently attained a part-time distance learning degree in social sciences (which he has not used for economic return), his income, wealth, social capital and so on mean that he remains within the dominated section of social space, even if relatively high within it – indeed, this is consonant with his background (upper-dominated parents, grammar school, unfinished first attempt at a degree) and work history (lower management) and represents inter-generational stasis. Finally, as Bourdieu and Boltanski (1981) indicated long ago, shared job titles, especially the vaguer ones attributed to parents' work, can camouflage a universe of difference in associated capital – 'engineer' is perhaps one of the most obvious examples of this semantic ambiguity, covering both highly skilled graduate-level professions and semi-skilled and routinized manual work (cf. Jackson and Marsden, 1962: 197), though 'computer programmer' may be a more contemporary example – and so, in designating capital levels here, knowledge of the precise *nature* of the occupation in question and general capital levels has been decisive.

Many of the interviews, especially those with members of the dominant class who were notably at ease with the world of higher education, were conducted at the university, but the rest took place at the interviewees' homes or workplaces, thus allowing access to the domains of their lifeworlds, the objects and people that populate them and, to some extent, the experiential parameters of their daily lives. There is a sense, therefore, in which a form of basic ethnographic research began before the interviews were under way, indeed before the houses or workplaces were even entered and the interviewees greeted as the areas in which people lived and worked revealed the material and cultural structuring of their lifeworlds and, what is more, the tastes characterizing their habitus. So, for example, the home of Samuel, an ambitious surgeon, was situated in a quiet suburban street comprised of large Georgian houses converted into luxury flats and, on the inside, was capacious and lavish with neat modern decoration, parquet flooring and expensive consumer goods, whereas Hannah's house in a predominantly red-brick street of varyingly decrepit semi-detacheds was much smaller, cluttered and filled with well-worn, functional furniture. Some homes, like Abby's, bore testament to the ample cultural capital of their inhabitants by having, in her case, poetry anthologies and works by Flaubert embedded in large literature-filled bookcases, whereas others, such as Yvonne's, contained few books but numerous family photos and knick-knacks instead (cf. Bourdieu, 1984: 379). Or again, compare the large bucolic cottage situated in a tranquil hamlet on a country lane belonging to Elizabeth, a computer programmer rich in both economic and cultural capital and cohabitating with her 'hospital-building' partner, with Dave's house wedged in the middle of a weathered terrace lining a narrow city street filled with ageing cars. Similarly with workplaces, some interviews were conducted in the plush, serene private offices of the interviewees (for example, Nigel and Oliver) while another took place in a cramped canteen smaller than either office but meant for a whole workforce (Phil, a low-level supervisor). Nevertheless, the exact biographical provenance of the various environs and artefacts and their place in systems of tastes and perceptions could not be illuminated until the interviews were under way.

The interviews themselves lasted eighty minutes on average, ranging from nearly three hours in length to just thirty minutes (one of only very few interviews to be below an hour in duration). Insofar as the interviews were opportunities for people to recount and explain their life choices and life paths, the interviewees shifted into an 'induced and accompanied self-analysis' of the manner described by Bourdieu in *The Weight of the World* (Bourdieu et al., 1999: 615), the specific substance of which is an empirical theme. In several cases though, and notably among those richer in cultural capital, such a self-analysis, whether or not it corresponded with the analysis of the sociologist, was accompanied by post-interview declarations that the experience had provoked contemplation and talk on themes and processes that they 'wouldn't usually think about' or that are 'taken for granted' (Isabelle), that it had prompted them to 'question themselves' and 'take stock' (Tessa) and even that it had been akin to a 'free therapy session' in which one can 'analyse, think about and discuss' oneself (Samuel). The language is, perhaps, reminiscent of that deployed by Beck and the others, but a distinction must be made here between *reflection* and *reflexivity*. The interviewees no doubt *reflected* on aspects of their lives usually 'taken for granted' in their descriptions and explorations, but this is not the *reflexivity* specifically described by Beck or Giddens: the interview situation, once entered, did not present a plurality of options in relation to life paths and lifestyles among which the interviewees then selected in a bid to answer the omnipresent question 'how shall I live?'; and it did not, so far as I could tell, induce a revision of an existing life plan or contribute to the formulation of a resolution on a sphere of life newly open to choice.

A note on generalization: From Bristol to the Western world?

And so we arrive at the final item on the roster: generalization. Unlike many studies deploying qualitative methods some level of extrapolation is desirable if the assessment of reflexivity is to move beyond the local level. Given the recurrent dissociation of qualitative research from any generalizing capacities, however, some may question the extent to which effective conclusions can be drawn on such sweeping theories on the basis of interviews with fifty-five residents of one city. It may help that the respondents were from Bristol, a fairly typical British (and Western) conurbation insofar as it has experienced a significant downturn in industrial activity and a corresponding burgeoning of its service sector (especially in finance, IT and low-level call centre work) in line with broader shifts in the economy and is starkly segregated spatially in terms of economic and cultural resources (for details, see Fenton and Dermott, 2006), and it may help that they were randomly sampled from the electoral register as a slice of the working population in general. But still, what inferences can one realistically make here?

Generalization to the national level is not actually too difficult. Insight can be taken from Bourdieu's claim, already hinted at in Chapter 3 when discussing the lifeworld, that the relational mode of thought in play here allows us to *overcome* the distinction between generalizing nomothetic research and localized ideographic analysis by allowing us to 'grasp particularity in generality and generality in particularity', with the end result that the unearthed themes can be treated, as Bachelard would say, as a 'particular case of the possible' (Bourdieu and Wacquant, 1992: 75). This guiding principle can, it seems to me, be clarified, elaborated and reinforced by two extant strategies of generalization in qualitative research. First of all, there are parallels with a neglected method of extrapolation dubbed 'theoretical generalization' in which wider inferences can be drawn from small-scale qualitative research on the basis of uncovering conceptualized structures and patterns with a broader applicability beyond the sample (Mitchell, 1983). If, for example, capital stocks are shown to be pivotal in shaping trajectories and habitus, then logically this can be generalized to some degree because capital stocks only exist in a relational structure, the social space, that is theoretically *trans-local* – a conceptualization also bolstered by nationwide statistical patterns. The consequences for the theories of reflexivity on this front are, therefore, far from restricted. At the same time, however, and especially in the case of more particular themes that emerge inductively (such as those regarding identity, subjectivity, typification bundles and so on), over-generalization must be avoided. As demanded by advocates of what has been called 'moderatum' generalization (Williams, 2000, 2002; Payne and Williams, 2005), therefore, the broader applicability of the inductive themes has to be only tentatively suggested, with a mind open to possible variations of greater and lesser shades given different regional, local and lifeworld conditions.

The study may still strike some as exclusively 'British' in its coverage, and indeed its conclusions can only be extended safely to the bounds of this nation state. But that does not mean the research has no relevance beyond those borders. Debates over reflexivity and individualization have gripped national sociological debates from Europe to the Antipodes, yet at the same time a steady stream of scholarship is constantly detailing the parameters and struggles of national social spaces in those same regions, and, given their shared political and educational cultures, this is not surprising.[6] Is it so unreasonable, then, to suppose that if the national structures are homologous then the explanation for and consequences of the persistence of social space unearthed in the UK could be an indicator of what, with national specificities, is going on in other Western nation states? If nothing else it would at least, I hope, encourage researchers to determine for themselves the applicability of the conclusions presented here in their homelands.

Of course the case of the Unites States is a little different. It is, after all, a nation where the concept of class has never had the same currency as in the Old World compared to ethnicity and race, and partly for this reason, and partly because it is rendered banal by the individualism that has always served as the general political culture, the reflexivity thesis and individualization have not had anywhere near the same impact (save perhaps through Giddens' Third Way, which appealed to Bill Clinton). But let there be no mistake: as Wacquant (2008b), Lareau (2000, 2003) and Lamont (1992, 2000) have demonstrated in their different and partial ways, there is a social space of difference in the US, there are inequalities of cultural and economic capital and there is a sensitivity to social difference, it is just that they are more closely entwined with ethno-racial domination in both structure and perception. Understanding the reproduction of these phenomena and refuting reflexivity and individualization might thus have more of an import than imagined, though less as a test of a particular theoretical perspective than as a possible counter to the political doxa that has ruled discourse in 'the land of the free' since its birth as an independent nation.

Notes

1. Introduction: From Affluence to Reflexivity

1. It is interesting to note, however, that some of the most influential explanatory accounts of the emergence of postmodern culture are either Marxist in orientation (for example, Harvey, 1989; Jameson, 1991) or inspired by Bourdieu's class theory (for example, Lash and Urry, 1987; Lash, 1990; Featherstone, 1991).
2. Social conditions also played their part, of course. In particular, it seems that the increased penetration of social science by women from self-proclaimed working-class backgrounds keen to document and make sense of their own experiences has been of consequence for the developments discussed below (see, for instance, Mahony and Zmroczek, 1997).
3. Though cultural class analysis began life as largely a British-based phenomenon, Bourdieu's class theory and its concerns have now swept global sociology, coming to prominence in the US, as evidenced by Weininger's (2005) essay in Wright's (2005) collection of 'approaches to class analysis' and in the research of Michelle Lamont (1992, 2000) and Annette Lareau (2000, 2003), and in other European nations and Australia (see contributions to Devine et al., 2005; also Bennett et al., 1999; Threadgold and Nilan, 2009).
4. To confer a name on one's standpoint is to stake a claim to existence and worth within the field of sociology, to demarcate one's position within it relative to others, and is thus inevitably an invidious and divisive task. It is, nevertheless, necessary to make clear the nature of the standpoint developed herein in order to avoid some of the more common charges levelled against Bourdieusian analysis. This is hardly straightforward, for each possible label triggers a cluster of assumptions and reactions when read through the schemes of perception in play in the field. 'Neo-Bourdieusianism', for example, would signify too great a conceptual break and distancing; 'structural phenomenology' obliterates the Bourdieusian starting point and sounds like a rather vague reworking of phenomenology; 'relational phenomenology', as used by McNay (2008), has a nice ring to it and manages to specify the approach to social structure, but, again, I am reluctant to lose the Bourdieusian reference. I have opted, in the end, for the rather awkward phrase 'phenomeno-Bourdieusianism' because it bears the fewest interpretational ambiguities – it is Bourdieusian at base, but emphasizes and expands the intersections with phenomenology.

2. Reflexivity and its Discontents

1. Beck's notion of reflexivity often seems to entail some amount of cogitation – agents now have to 'think, calculate, plan, adjust, negotiate, define, revoke' rather than rely on habitualization or routine (Beck and Beck-Gernsheim, 2002:6) – but it can also be more 'spontaneous' in form and thus differs from Archer's (2007a) conception of reflexivity, detailed later, which specifically denotes considerable mental deliberation (cf. Lash, 2002).

2. Bauman (2008: 83) has recently stated that the metaphor of a ship continuously raising and lowering anchor at different ports is more apt than disembedding or uprooting, but the essential point is the same.

3. Bauman does not actually use the term reflexivity itself, and is critical of the notion of 'reflexive modernity', but it is there in all but name.

4. Fateful moments are essentially a brand of what Giddens earlier referred to as 'critical situations', that is, situations in which routine is radically disrupted and ontological security threatened (see Giddens, 1979: 124; 1984: 60–1).

5. Witness, for example, Goldthorpe's (2007b: 285n14) claim that it is *irrelevant* (or 'of no consequence') whether choices take the form of explicit decision-making procedures or are implicit, piecemeal and emergent over time (cf. also his bold assertion that instrumental rationality is a human universal and therefore not, as some studies have claimed, differentially distributed – 2007a: 177–83; 2007b: 84).

6. To be fair to Goldthorpe, it should be noted that he does allow room for other explanations of regularities so long as they can prove themselves empirically (which he claims they cannot) and admits that qualitative methods, specifically ethnography, are better equipped to clarify causal processes 'on the ground' and should thus be used as a complement to survey research (2007a: 81ff) – though neither he nor any of his associates actually use such methods.

7. Two of the studies cited here focus only on youth transitions, an apt but not comprehensive frame for assessing individualization, while Nollmann and Strasser use quantitative methods to study the subjective dimension and thus fall victim to the same problems as Goldthorpe.

8. Savage et al. (2001: 885–7) do distinguish those who, they claim, can 'reflexively' play with class labels given their possession of ample cultural capital from those who are more defensive – but here 'reflexivity' is shorn of its connections with Giddens and Beck and reduced to simply displaying more familiarity with the discourse of class.

9. To be more precise, the sample is drawn from four residential areas illustrative of differing middle-class fractions, some of which are deemed to include the 'upper', 'older' or 'affluent' working class in their midst (Savage, 2000: 119n14; Savage et al., 2001: 878–9), and if one looks to the details given by Savage et al. (2005a: 209–14) it can be seen that though occasionally cleaners, builders, shop assistants and more do find their way into the sample there is an overwhelming predominance of professional–managerial occupations. Savage's (2005, 2007) more recent secondary analyses of older data sets do not remedy this situation: his use of Mass Observation data suffers precisely the same problem while in his reinterpretation of the *Affluent Worker* interviews the time difference between the two samples undercuts any real claim to comparability.

3. Conceptualizing Class and Reconceptualizing Reflexivity

1. Expectations and aspirations are, however, fair empirical indicators of the dispositions of the habitus, as an orientation to present and future action

adapted to objective probabilities, which is itself an indicator of the cultural (and economic) capital marking an agent's situation. Devine's misunderstanding, insofar as she neglects the habitus, is therefore essentially a short circuit.

2. Economic and cultural capital gain their meaning and symbolic value within a national culture, educational system, state-sanctioned legal structure and economic climate and so the social space is, therefore, a national-level concept, even if the impact of global processes, such as immigration, the international economic field or cultural imperialism, must be recognized. This, as I have argued elsewhere (Atkinson, 2007b, 2007c), does not weaken class analysis in the way Beck claims.

3. Originally Bourdieu inserted the caveat that learning took place through practice rather than discourse only where education had not been institutionalized, but since then, particularly in *Pascalian Meditations* and Wacquant's *Body and Soul*, the proposition seems to have been generalized.

4. In phenomenology this distinction is usually articulated using the notions of 'theme' and 'horizon' (for example, Gurwitsch, 1964; Schutz, 1970b; Husserl, 1976), but given the multiple definitions and uses of them by different phenomenologists I think Searle's alternative distinction between core and periphery probably allows greater clarity.

5. Commitment need not be firm, but is what distinguishes projected action proper from projects *in potentia* – that is, courses of action considered but not then committed to. A distinction also needs to be made between the consideration of likely or practical courses of action and unrealistic fantasies, the latter of which sink into the spontaneous activity of mere thinking (Schutz, 1962: 212).

6. This avoids Schutz's inability, premised on his purely mental and deliberative depiction of projection, to conceive of brief or intersubjectively determined projects and to thus keep spontaneous action and projection clearly distinguishable (see Moerman, 1988: 65ff; Crossley, 1996: 81).

7. There is also the possibility, raised very infrequently and only suggestively by Bourdieu, of the effect of relational differences that are not organized as fields. He speaks, for example, of various 'spaces' or 'universes' having some, but not all, of the features of fields, such as world sociology (Bourdieu, 1991b), and his thoughts on gender probably fit here too (Bourdieu, 2001). This is all in need of further development, but will not concern us here.

8. Bennett et al. (2009), for their part, follow suit and also describe and reject a crude image of a 'unified habitus', as if all members of an analytical class were identical rather than internally dispersed and united only by varying degrees of family resemblance, despite the fact that they claim to champion a relational approach like Bourdieu. Then again, what they actually mean by the latter assertion beyond using correspondence analysis and the language of cultural fields is unclear, especially since they discard the notion of social space from the outset. They also, like Lahire, pay insufficient attention to trajectory in Bourdieu's model, but they have their own problems too. There is, for example, an unquestioned use of Goldthorpe-based occupational classifications and a constricted understanding of cultural capital as the possession of a disinterested Kantian aesthetic disposition, as mentioned earlier, but also a separation of 'class' from education level, a dismissal of the idea

that even apparently non-class-related practices can be seen in class terms (as 'slumming it', for example) for no good reason and an unwarranted appeal to modish tropes of 'fluidity' and 'multiple subjectivities' and the banalities of actor–network theory. No doubt stemming in part from the fact that the book is written by a collection of scholars known to have very different views of Bourdieu's value, and whatever the merits of their substantive findings, their confident claims that the patterns of cultural consumption unveiled in the UK conclusively demonstrate that much of Bourdieu's theory is outdated or useless are thus questionable to say the least.

9. This individual-level reading of the lifeworld, defended at greater length elsewhere (Atkinson, 2010), is opposed to the more common collective one proposed by most phenomenologists, and is drawn from Schutz (1970a: 320) and Natanson (1974: chap. 13). See Steinbock (1995: 174ff) for a discussion of the different levels of conception.

10. For concrete studies of this 'class making' process, see Boltanski's (1984, 1987) work on the construction of 'cadres' as a class in France and Wacquant's (1991) discussion of Jurgen Kocka's analysis of the rise of the German *Angestellten*.

11. Wacquant also describes the subjective dimension of the habitus as 'definitions of the situation, typifications, interpretive procedures' (in Bourdieu and Wacquant, 1992: 12), but nowhere does Bourdieu corroborate this interpretation.

12. It should be noted at this point that some practices and objects, such as the Muslim hijab, are typified and paired according to ethnicity rather than position in social space, though these inevitably appear in combination with other practices, goods and modes of being that do signify the latter. Furthermore, some practices and goods – for example, 'gangsta rap' music – are typified with and thus signify both ethnicity *and* position in social space at the same time (in this case black and dominated). Either way, discrimination based on the reading of ethnic or ethnically flavoured signs (or for that matter skin colour), bound up with the symbolic and discursive construction of the ethnic 'group', impacts upon the agent's position in social space and their habitus in the way suggested. This applies *mutatis mutandis* to gender as well.

13. In more recent work Giddens (2007: 66) has claimed that lifestyle differences between classes are based not on 'financial constraints' (that is, economic capital) but on different tastes, without recognizing that different tastes are formed on the basis of the structuring of lifeworlds by economic capital.

4. Educational Reproduction Today

1. For more information on the interviewees' structural locations, see the Appendix.

2. Sullivan's (2007) separation of participation in the 'beaux arts' from the other ingredients of passive transmission may be useful for quantitative clarity, but makes less sense in the observation of lived reality. Insofar as it is not consciously related to the specific demands of the school, is woven into the taken-for-granted experiences and routines of the lifeworld and, above

all, inculcates not simply a hermetically sealed knowledge of art, theatre or whatever in which familiarity allows recognition and the drawing of connections, but the same *general* symbolic mastery as the other elements – allegory, critique, abstraction (what is 'beauty') and so on – the former can still be considered a species of passive transmission and is thus thrown into the mixture here.

3. The terms 'institutional habitus' and 'familial habitus' have been developed by Reay et al. (2005) to capture the expectations of educational institutions and families respectively, but I have serious reservations apropos these terms because they not only extend to the aggregate and substantialist level an individual and relational property but threaten to smother the complex of complementary and contradictory expectations conveyed by individuals, who after all have their own habitus and trajectories, within these domains of the lifeworld. The term 'ethos', as a school-specific doxa, seems better able to capture what Reay et al. intend to illuminate.

4. There are only a couple of deviations from this, representing 'disjointed trajectories', but there is not the space to consider them here.

5. Joe believes he has now 'grown out of' his dyslexia insofar as the problems he had with reading and writing have now been completely overcome. Given that dyslexia is not actually a condition people 'grow out of', this raises the question of whether his diagnosis was in fact a *misdiagnosis* of an acute lack of cultural capital, and hence a naturalization of educational inequality.

6. There are obvious intersections with the effects of gender here as well: Yvonne was expected to be a secretary rather than, say, a bricklayer, because 'that's what *girls* did'.

7. I deal here only with those upwardly mobile through education; the select few mobile in the course of their occupational trajectories are considered in Chapter 5.

8. This is, then, a long-standing class-based disposition that, instead of primarily encouraging saving and fertility strategies, as for Bourdieu's petite bourgeoisie (which included draughtsmen, like Lisa's father), realizes itself through the expanded education system and promotion.

5. Topographical Trajectories

1. In the case of objectified or institutionalized cultural capital – for example, having the 'right' qualification – open doors are harder to observe than closed ones given that interviewees only applied for jobs in line with their credentials, other than perhaps the instances where interviewees have attended conversion courses or postgraduate studies, sometimes later in life, and the degrees they possess act as their entry tickets, as is the case for Karen (from psychology at Oxford to graduate entry medicine), Lisa (from a degree and PGCE to a HR diploma) or Adrian (from dentistry to law).

2. All this explains, but heavily qualifies, the research approvingly cited by the UK Cabinet Office (2008) claiming that possession of diverse social networks, which Archer would see as offering a form of contextual discontinuity, is conducive to higher aspirations among otherwise disadvantaged individuals.

3. There are a handful of cases where there seems to be more of a break with the occupational past, but these are easily explicable by two factors appearing either on their own or in tandem: social capital, as people are inducted into forms of work with which they are familiar through family or friends (as in Lisa's move into HR from teaching, or Gina and Gillian's moves into teaching support from work as a mechanic and office worker respectively), and the jobs departed and taken up requiring few specific skills meaning movement between them is 'easy and typical', to use Weber's phrase (as in the case of Zoe's move from routine sales work to low-level office work). It should be remembered that just because these interviewees have left their previous lines of work, it does not mean that they have not left indelible marks on their habitus in some way – Gina, for example, continues to repair vehicles for friends and family.

4. It is enlightening to compare Eddie with Wendy, a deputy head teacher upwardly mobile from her origins. She shared a remarkably similar semi-rural life and commitment to sports and outdoor pursuits through her youth, but because she had slightly more advantages inhering in her situation – her father was a skilled avionics electronics system designer while her mother was a sales representative-turned-market researcher who, though still dominated overall, were notably more affluent and education-focused than Eddie's parents – she could draw on their resources, and their discourse of self-betterment, to achieve, with a dose of struggle, social ascent.

6.   Distinction and Denigration

1. Statistics show that gardening and DIY are differentiated according to ONS class categories at the national level (Roberts, 2004: 60), but reliable data on cooking as a leisure pursuit is harder to come by.

2. Of course, even where it is mentioned by the dominated in surveys the divergent *meanings* of 'gardening' attached to separate zones of social space (as aesthetic project or functional maintenance) must be borne in mind. So too must the complicating factor of age, for as dominated agents grow older and their children leave home the functional demands placed on gardens lessen and, especially when coupled with the free time that comes with retirement, they can become sites of leisure. Thus Maureen, her grown son having moved out some time ago, was one of only two dominated interviewees to mention gardening as a leisure pursuit, but even she admitted that time pressures had resulted in neglect and an overgrown lawn.

3. Chan and Goldthorpe (2007b: 10–11) draw on old psychological theories to argue that the educated have a higher 'information-processing' capacity, which requires more 'complex stimuli' for pleasure, without recognising that such a capacity is socially endowed as cultural capital and, therefore, definitive of class and not, as they claim, status.

4. Though the prevalence of the distinction in the general population is not able to be questioned here, the difference found by Bennett et al. (2009) between outwardly and inwardly oriented practices corresponding to gender – in the case of reading, between factual or science fiction books (male) and

romance or human interest books (female) – was not particularly evident among the interviewees.

5. In Roberts' (2004: 61) data skittles appears to be practised more by non-manual workers than manual workers, but this is doubtless because it is conflated with ten-pin bowling, an activity which, though similar in technical structure, is wholly different in cultural presentation (one need only think of bowling alleys in large entertainment complexes and malls).

6. Labels hide a world of internal difference. Phil's boxercise classes seem to be particularly close to boxing *per se*, involving sparring but not fighting, but other classes going under the same name, in different gyms in different locales, are more of a synchronized keep-fit class and no doubt attract a rather different clientele.

7. These two figures were also cited by Barry, who was extremely self-conscious of his lack of inherited cultural capital.

8. In most cases this was unprompted, but sometimes it flowed forth after the topic of class had been explicitly broached. Where explicit class labels are used that should be borne in mind.

9. A number of interviewees reported instances where, in the company of others from different parts of social space, they had attempted or would attempt to 'adapt' their behaviour by, for example, altering accents, pronunciation and the content of conversation.

7. 'Class' as Discursive and Political Construct

1. The respondents were asked towards the end of the interview the following question: 'when I say social class, what does that mean to you/make you think of?' A series of follow-up questions were then pursued.

2. The corollary of this, of course, is that the category of 'snob' is highly stigmatized, leading several interviewees to overtly deny or worry they were being 'snobby' when making a comment, or to recount instances where they have, unjustifiably and disagreeably in their view, been perceived as 'snobs'.

3. Savage et al. (2001: 887) dub this the 'omnivoric refrain' in reference to the literature on omnivorousness and view it as part of a strategy by both dominated *and* dominant agents to establish one's 'ordinariness' and 'normality'. However, though some of the dominant interviewees here did also occasionally claim to want to, or be able to, 'treat everyone the same' – a classic phrase in strategies of condescension (Bourdieu, 1991a; cf. Sayer, 2005: 171–6) – it was not in the same defensive context or tied to any claim to normality.

4. The fact that both Jimmy and Phil were in or considering supervisory roles, and Eddie's political persuasion seen above, must be taken into account when making sense of their ambivalence.

5. It should be made clear that though Bristol, like all cities, has had its specificities in its trade union history (for example, the influence of the local left-wing MP Tony Benn), which have acted to refract the manifestations of these processes, it remains entirely typical of national patterns and developments (see Kelly and Richardson, 1996).

8. Conclusion: Rigid Relations through Shifting Substance

1. See, for example, the contributions to Silva and Warde's (2010) assessment of Bourdieu's legacy, in which various representatives of class research are categorized as at best 'partial appropriators' but also 'critical revisers' or 'repudiators'.
2. All quotations are taken from the Conservatives' 1992 General Election Manifesto, archived online by the Political Sciences Resources service of Keele University at http://www.psr.keele.ac.uk/area/uk/man/ con92.htm#all.
3. Nor does it look likely to get much better in the near future given that, at the time of writing, the Conservatives, whose politics have barely even superficially changed, have just formed a new coalition government with the Liberal Democrats and delivered an ominous Budget.
4. All quotes are taken from a speech on social mobility delivered by Gordon Brown to educational professionals on 23 June 2008, archived online by the Prime Minister's Office: http://www.number10.gov.uk/Page16181. They could just have easily have been lifted from the more recent report into social mobility produced under the guidance of former Minister for Health Alan Milburn (Panel on Fair Access to the Professions, 2009).
5. The focus here is on education and social mobility, but current employment policies are problematic too. Lifelong learning and reskilling could be positive facets of occupational life, for example, if only the range of options were not narrowed by economic or cultural barriers and, more importantly, *forced* upon individuals by the caprices of *laissez-faire* capitalism.

Appendix: The Search Process

1. There is, in fact, no extant phenomenological tradition of research into the lifeworld in sociology – indeed, phenomenological sociology as a whole was largely subsumed into ethnomethodology in the 1960s and 1970s – and so this understanding is drawn from theoretical sources and from the disciplines of psychology, education and nursing research, where phenomenology remains a strong school of thought in its own right (see, for example, Moustakas, 1994; Creswell, 1998; Ashworth, 2003; Titchen and Hobson, 2005). The latter are, however, usually less focused on the lifeworld than on the experience of specific phenomena and, furthermore, where it is investigated it is often understood in an overly idealist manner.
2. The language here is Kantian, and indeed there are links with the Kant-inspired transcendental argument of critical realism – what must be the case *a priori* for this doxic experience to have the features it does and to be different from or similar to the doxic experience of others? The answer: social structures. See Bhaskar (1975, 1998), Porter (2002) on critical realism and social structures in qualitative research, and Wacquant (1989), Vandenberghe (1999) and Potter (2000) on the ambiguous relationship between Bourdieu and critical realism.
3. The only exception is Hannah, who moved from full-time to part-time work between initial contact and the interview date.
4. I should make clear that, while I have tried to be sensitive to the intersecting effects of gender throughout the analysis and mention them where salient,

the defining task of the study necessitates a focus on the forces emanating from the structures of social space and uniting experience across gender differences. So, for example, the differential possession of symbolic mastery plays out very differently according to gender – among the dominated, the relative lack manifesting in an interest in manual skills or physical toughness versus family-centredness and a focus on 'glamorous' beauty (Archer et al, 2007b); among the dominant, differentiating the avenues of further learning in which symbolic mastery is applied, such as 'hard' natural sciences and philosophy versus 'soft' arts and social sciences (cf. Bourdieu and Passeron, 1979). Indeed, its very genesis is partially shaped by relatively autonomous gender dispositions, such as the higher valuation of educational achievement among girls given different perceptions of its place within their lives produced by expectations of women's work–family commitments (cf. Mikelson, 2003), the 'defence' of quiet determination noted by Walkerdine et al. (1999), which Bourdieu (2001) himself would describe much more contentiously as 'docility', and the fact that mothers are disproportionately engaged in childrearing (Reay, 1998a). Nevertheless, in order to respond to Beck and the others it is, in this instance, the role of cultural capital and its inter-generational transmission *per se* that are emphasized.

5. The relational terms dominant and dominated class are used to demarcate structural positions rather than 'working class' and so on in order to maintain analytical rigour and break with lay categorizations and constructions that would cloud analysis. Those classified as dominant outnumber the dominated on a ratio of 3:2, though when the upwardly mobile are treated as a separate third category the numbers are more even. Nevertheless it should be borne in mind that the processes revealed among the dominant have the greater weight of empirical support.

6. See, for example, Sintas and Álvarez (2002) on Spain, Vester (2005) on Germany, Prieur et al. (2008) on Denmark and Bennett et al. (1999) on Australia.

References

Adams, M. (2006), 'Hybridizing Habitus and Reflexivity: Towards an Understanding of Contemporary Identity?', *Sociology*, 40 (3): 511–28.

Adams, M. (2008), *The Reflexive Self: A Critical Assessment of Giddens's Theory of Self-Identity*. VDM: Sarrbrücken.

Adkins, L. (2003), 'Reflexivity: Freedom or Habit of Gender?', *Theory, Culture & Society*, 20 (6): 21–42.

Adkins, L. and Skeggs, B. (eds) (2004), *Feminism After Bourdieu*. Oxford: Blackwell.

Adonis, A. and Pollard, S. (1997), *A Class Act: The Myth of Britain's Classless Society*. London: Penguin.

Allatt, P. (1993), 'Becoming Privileged: The Role of Family Processes', in I. Bates and G. Riseborough (eds), *Youth and Inequality*. Buckingham: Open University Press, 139–59.

Alt, J. (1976), 'Beyond Class: The Decline of Industrial Labor and Leisure', *Telos*, 28: 55–80.

Anscombe, G. E. M. (1957), *Intention*. Oxford: Blackwell.

Anthias, F. (1999), 'Theorising Identity, Difference and Social Divisions', in M. O'Brien, S. Penna and C. Hay (eds), *Theorising Modernity: Reflexivity, Environment and Identity in Giddens' Social Theory*. London: Longman, 156–78.

Archer, L., Hollingworth, S. and Halsall, A. (2007a), 'University's Not for Me – I'm a Nike Person: Urban, Working-Class Young People's Negotiations of "Style", Identity and Educational Engagement', *Sociology*, 41 (2): 219–37.

Archer, L., Halsall, A. and Hollingworth, S. (2007b), 'Class, Gender, (Hetero)sexuality and Schooling: Paradoxes Within Working-Class Girls' Engagement with Education and Post-16 Aspirations', *British Journal of Sociology of Education* 28 (2): 165–80.

Archer, L., Hutchings, M. and Ross, A. (2003), *Higher Education and Social Class: Issues of Exclusion and Inclusion*. London: RoutledgeFalmer.

Archer, M. S. (2000), *Being Human: The Problem of Agency*. Cambridge: Cambridge University Press.

Archer, M. S. (2003), *Structure, Agency and the Internal Conversation*. Cambridge: Cambridge University Press.

Archer, M. S. (2007a), *Making our Way through the World: Human Reflexivity and Social Mobility*. Cambridge: Cambridge University Press.

Archer, M. S. (2007b) 'The Trajectory of the Morphogenetic Approach' *Sociologia, Problemas e Práticas*, 54: 35–47.

Ashworth, P. (2003), 'An Approach to Phenomenological Psychology: The Contingencies of the Lifeworld', *Journal of Phenomenological Psychology*, 34 (2): 145–56.

Atkinson, W. (2007a), 'Anthony Giddens as Adversary of Class Analysis', *Sociology*, 41 (3): 533–49.

Atkinson, W. (2007b), 'Beck, Individualization and the Death of Class: A Critique', *The British Journal of Sociology*, 58 (3): 349–66.

Atkinson, W. (2007c), 'Beyond False Oppositions: A Reply to Beck', *British Journal of Sociology*, 58 (4): 707–15.

Atkinson, W. (2008), 'Not All That is Solid has Melted into Air (or Liquid)', *Sociological Review*, 56 (1): 1–17.

Atkinson, W. (2009), 'Rethinking the Work-Class Nexus: Theoretical Foundations for Recent Trends', *Sociology*, 43 (5): 896–912.

Atkinson, W. (2010), 'Phenomenological Additions to the Bourdieusian Toolbox: Two Problems for Bourdieu, Two Solutions from Schutz', *Sociological Theory*, 28 (1): 1–19.

Ball, S. J. (2003), *Class Strategies and the Education Market: The Middle Classes and Social Advantage*. London: RoutledgeFalmer.

Ball, S. J. (2008), *The Education Debate*. Bristol: Policy Press.

Ball, S. J., Maguire, M. and Macrae, S. (2000), *Choice, Pathways and Transitions Post-16: New Youth, New Economies in the Global City*. London: RoutledgeFalmer.

Baudrillard, J. (2001), *Selected Writings* (2nd edn). Cambridge: Polity Press.

Bauman, Z. (1982), *Memories of Class: The Pre-History and After-Life of Class*. London: Routledge and Kegan Paul.

Bauman, Z. (1987), *Legislators and Interpreters: On Modernity, Post-Modernity and Intellectuals*. Cambridge: Polity Press.

Bauman, Z. (1988), *Freedom*. Milton Keynes: Open University Press.

Bauman, Z. (1998a), *Work, Consumerism and the New Poor*. Buckingham: Open University Press.

Bauman, Z. (1998b), *Globalization: The Human Consequences*. Cambridge: Polity Press.

Bauman, Z. (1999), *In Search of Politics*. Cambridge: Polity Press.

Bauman, Z. (2000), *Liquid Modernity*. Cambridge: Polity Press.

Bauman, Z. (2001), *The Individualized Society*. Cambridge: Polity Press.

Bauman, Z. (2002), *Society Under Siege*. Cambridge: Polity Press.

Bauman, Z. (2004a), *Identity: Conversations with Benedetto Vecchi*. Cambridge: Polity Press.

Bauman, Z. (2004b), *Wasted Lives*. Cambridge: Polity Press.

Bauman, Z. (2005), *Liquid Life*. Cambridge: Polity Press.

Bauman, Z. (2007a), *Liquid Times: Living in an Age of Uncertainty*. Cambridge: Polity Press.

Bauman, Z. (2007b), *Consuming Life*. Cambridge: Polity Press.

Bauman, Z. (2008), *Art of Life*. Cambridge: Polity Press.

Bell, D. (1976), *The Cultural Contradictions of Capitalism*. London: Heinemann.

Beck, U. (1992), *Risk Society: Toward a New Modernity*. London: Sage.

Beck, U. (1994), 'The Reinvention of Politics: Towards a Theory of Reflexive Modernization', in U. Beck, A. Giddens and S. Lash, *Reflexive Modernization: Politics, Tradition and Aesthetics in the Modern Order*. Cambridge: Polity Press, 1–54.

Beck, U. (1997), *The Reinvention of Politics: Rethinking Modernity in the Global Social Order*. Cambridge: Polity Press.

Beck, U. (1998), *Democracy Without Enemies*. Cambridge: Polity Press.

Beck, U. (2000a), *The Brave New World of Work*. Cambridge: Polity Press.

Beck, U. (2000b), *What is Globalization?* Cambridge: Polity Press.

Beck, U. (2002), 'The Cosmopolitan Society and its Enemies', *Theory, Culture & Society*, 19 (1–2): 17–44.

Beck, U. (2007), 'Beyond Class and Nation: Reframing Social Inequalities in a Globalizing World', *The British Journal of Sociology*, 58 (4): 679–705.

Beck, U. (2009), *World at Risk*. Cambridge: Polity Press.

Beck, U. and Beck-Gernsheim, E. (2002), *Individualization: Institutionalized Individualism and its Social and Political Consequences*. London: Sage.

Beck, U. and Lau, C. (2005), 'Second Modernity as a Research Agenda: Theoretical and Empirical Explorations in the "Meta-Change" of Modern Society', *The British Journal of Sociology*, 56 (4): 525–57.

Beck, U. and Willms, J. (2004), *Conversations with Ulrich Beck*. Cambridge: Polity Press.

Beck, U., Bonss, W. and Lau, C. (2003), 'The Theory of Reflexive Modernization: Problematic, Hypotheses and Research Programme', *Theory, Culture & Society*, 20 (2): 1–33.

Bennett, A. (1999), 'Subcultures or Neo-Tribes? Rethinking the Relationship between Youth, Style and Musical Taste', *Sociology*, 33 (3): 599–617.

Bennett, T. (2007), 'Habitus *Clivé*: Aesthetics and Politics in the Work of Pierre Bourdieu', *New Literary History*, 38: 201–28.

Bennett, T., Emmison, M. and Frow, J. (1999), *Accounting for Taste: Australian Everyday Cultures*. Cambridge: Cambridge University Press.

Bennett, T., Savage, M., Silva, E. B., Warde, A., Gayo-Cal, M. and Wright, D. (2009), *Culture, Class, Distinction*. London: Routledge.

Berger, P. L. (1977), *Facing Up to Modernity: Excursions in Society, Politics, and Religion*. New York: Basic Books.

Berger, P. L. and Luckmann, T. (1967), *The Social Construction of Reality: A Treatise in the Sociology of Knowledge*. London: Penguin.

Berger, P. L., Berger, B. and Kellner, H. (1974), *The Homeless Mind: Modernization and Consciousness*. Harmondsworth: Penguin.

Beynon, H. (1984), *Working for Ford* (2nd edn). Harmondsworth: Penguin.

Bhaskar, R. (1975), *A Realist Theory of Science*. Leeds: Leeds Books.

Bhaskar, R. (1998), *The Possibility of Naturalism* (3rd edn). London: Routledge.

Boltanski, L. (1984), 'How a Social Group Objectified Itself: "Cadres" in France, 1936–45', *Social Science Information*, 23 (3): 469–91.

Boltanski, L. (1987), *The Making of a Class: Cadres in French Society*. Cambridge: Cambridge University Press.

Boltanski, L. and Chiapello, È. (2005), *The New Spirit of Capitalism*. London: Verso.

Bottero, W. (2004), 'Class Identities and the Identity of Class', *Sociology*, 38 (5): 985–1003.

Boudon, R. (1998), 'Social Mechanisms without Black Boxes', in P. Hedström and R. Swedberg (eds), *Social Mechanisms*. Cambridge: Cambridge University Press, 172–203.

Bourdieu, P. (1977), *Outline of a Theory of Practice*. Cambridge: Cambridge University Press.

Bourdieu, P. (1979), *Algeria 1960*. Cambridge: Cambridge University Press.

Bourdieu, P. (1981), 'Men and Machines', in K. Knorr-Cetina and A. V. Cicourel (eds), *Advances in Social Theory and Methodology: Toward an Integration of Micro- and Macro-Sociologies*. London: Routledge and Kegan Paul, 304–17.

Bourdieu, P. (1984), *Distinction: A Social Critique of the Judgement of Taste*. London: Routledge.

Bourdieu, P. (1987), 'What Makes a Social Class? On the Theoretical and Practical Existence of Groups', *Berkeley Journal of Sociology*, 32: 1–17.

Bourdieu, P. (1988), *Homo Academicus*. Cambridge: Polity Press.

Bourdieu, P. (1990a), *The Logic of Practice*. Cambridge: Polity Press.

Bourdieu, P. (1990b), *In Other Words: Essays Towards a Reflexive Sociology*. Cambridge: Polity Press.

Bourdieu, P. (1991a), *Language and Symbolic Power*. Cambridge: Polity Press.

Bourdieu, P. (1991b), 'Epilogue: On the Possibility of a Field of World Sociology', in P. Bourdieu and J. Coleman (eds), *Social Theory for a Changing Society*. Boulder, Col.: Westview Press, 373–87.

Bourdieu, P. (1993a), *Sociology in Question*. London: Sage.

Bourdieu, P. (1993b), 'Concluding Remarks: For a Sociogenetic Understanding of Intellectual Works', in C. Calhoun, E. LiPuma and M. Postone, (eds), *Bourdieu: Critical Perspectives*. Cambridge: Polity Press, 263–75.

Bourdieu, P. (1996a), *The State Nobility: Elite Schools in the Field of Power*. Cambridge: Polity Press.

Bourdieu, P. (1996b), *The Rules of Art: Genesis and Structure of the Literary Field*. Cambridge: Polity Press.

Bourdieu, P. (1997a), 'The Forms of Capital', in A. H. Halsey, H. Lauder, P. Brown and A. S. Wells (eds), *Education: Culture, Economy, and Society*. Oxford: Oxford University Press, 46–58.

Bourdieu, P. (1997b), 'Passport to Duke', *Metaphilsophy*, 28 (4): 449–55.

Bourdieu, P. (1998a), *Practical Reason: On the Theory of Action*. Cambridge: Polity Press.

Bourdieu, P. (1998b), *On Television and Journalism*. London: Pluto Press.

Bourdieu, P. (1998c), *Acts of Resistance*. Cambridge: Polity Press.

Bourdieu, P. (1999), 'The Social Conditions of the International Circulation of Ideas' in R. Shusterman (ed.), *Bourdieu: A Critical Reader*. Cambridge: Blackwell, pp. 220–8.

Bourdieu, P. (2000a), *Pascalian Meditations*. Cambridge: Polity Press.

Bourdieu, P. (2000b), 'The Biographical Illusion', in P. du Gay, J. Evans and P. Redman (eds), *Identity: A Reader*. London: Sage, 297–303.

Bourdieu, P. (2001), *Masculine Domination*. Cambridge: Polity Press.

Bourdieu, P. (2005), *The Social Structures of the Economy*. Cambridge: Polity Press.

Bourdieu, P. (2007), *Sketch for a Self-Analysis*. Cambridge: Polity Press.

Bourdieu, P. and Boltanski, L. (1981) 'The Educational System and the Economy: Titles and Jobs' in C. C. Lemert (Ed.) *French Sociology: Rupture and Renewal Since 1968*. New York: Columbia University Press, 141–51.

Bourdieu, P. and Passeron, J.-C. (1979), *The Inheritors: French Students and Their Relation to Culture*. Chicago: University of Chicago Press.

Bourdieu, P. and Passeron, J.-C. (1990), *Reproduction in Education, Society and Culture* (2nd edn). London: Sage.

Bourdieu, P. and Wacquant, L. J. D. (1992), *An Invitation to Reflexive Sociology*. Cambridge: Polity Press.

Bourdieu, P. and Wacquant, L. J. D. (2001), 'NewLiberalSpeak: Notes on the New Planetary Vulgate', *Radical Philosophy*, 105: 2–5.

Bourdieu, P., Chamboredon, J.-C. and Passeron, J.-C. (1991a), *The Craft of Sociology: Epistemological Preliminaries*. New York: Walter de Gruyter.

Bourdieu, P., Darbel, A and Schnapper, D. (1991b), *The Love of Art: European Art Museums and their Public*. Cambridge: Polity.

Bourdieu, P., Boltanski, L., Castel, R., Chamboredon, J.-C. and Schnapper, D. (1990), *Photography: A Middle-Brow Art*. Cambridge: Polity.

Bourdieu, P. et al. (1999), *The Weight of the World: Social Suffering in Contemporary Society*. Stanford: Stanford University Press.

Bradley, H. (2002), *Gender and Power in the Workplace*. London: Macmillan.

Bradley, H., Erickson, M., Stephenson, C. and Williams, S. (2000), *Myths at Work*. Cambridge: Polity Press.

Brannen, J. and Nilsen, A. (2005), 'Individualisation, Choice and Structure: A Discussion of Current Trends in Sociological Analysis', *The Sociological Review*, 53 (3): 412–28.

Breen, R. and Rottman, D. B. (1995), *Class Stratification: A Comparative Perspective*. Hemel Hempstead: Harvester Wheatsheaf.

Brown, P. and Lauder, H. (1996), 'Education, Globalization, and Economic Development', *Journal of Educational Policy*, 11: 1–24.

Brückner, H. and Mayer, K. U. (2005), 'De-standardization of the Life-Course', *Advances in Life Course Research*, 9: 27–53.

Bufton, S. (2003), 'The Lifeworld of the University Student: Habitus and Social Class', *Journal of Phenomenological Psychology*, 34 (2): 207–34.

Burawoy, M. (1979), *Manufacturing Consent*. Chicago: university of Chicago Press.

Cabinet Office (2008), *Aspiration and Attainment Amongst Young People in Deprived Communities*. London: Cabinet Office.

Cannadine, D. (1998), *Class in Britain*. London: Penguin.

Cassirer, E. (1923), *Substance and Function and Einstein's Theory of Relativity*. Chicago: Open Court.

Chan, T. W. and Goldthorpe, J. H. (2004), 'Is There a Status Order in Contemporary British Society? Evidence from the Occupational Structure of Friendship', *European Sociological Review*, 20 (5): 383–401.

Chan, T. W. and Goldthorpe, J. H. (2005), 'The Social Stratification of Theatre, Dance and Cinema Attendance', *Cultural Trends*, 14 (3): 193–212.

Chan, T. W. and Goldthorpe, J. H. (2007a), 'Class and Status: The Conceptual Distinction and its Empirical Relevance', *American Sociological Review*, 72: 512–32.

Chan, T. W. and Goldthorpe, J. H. (2007b), 'Social Stratification and Cultural Consumption: Music in England', *European Sociological Review*, 23 (1): 1–19.

Chan, T. W. and Goldthorpe, J. H. (2007c), 'Social Status and Newspaper Readership', *The American Journal of Sociology*, 112 (4): 1095–134.

Chan, T. W. and Goldthorpe, J. H. (2007d), 'Social Stratification and Cultural Consumption: The Visual Arts in England', *Poetics*, 35: 168–90.

Chan, T. W. and Goldthorpe, J. H. (2007e), 'Data, Methods and Interpretation in Analyses of Cultural Consumption: A Reply to Peterson and Wuggenig', *Poetics*, 35: 317–29.

Changeux, J.-P. (1985), *Neuronal Man*. New York: Pantheon.

Charlesworth, S. J. (2000), *A Phenomenology of Working Class Experience*. Cambridge: Cambridge University Press.

Clark, T. N. (2001), 'The Debate over "Are Social Classes Dying?"', in T. N. Clark and S. M. Lipset (eds), *The Breakdown of Class Politics: A Debate on Post-Industrial Stratification*. Baltimore: John Hopkins University Press, 273–319.

Clark, T. N. and Lipset, S. M. (1991), 'Are Social Classes Dying?', *International Sociology*, 6 (4): 397–410.

Clark, T. N. and Lipset, S. M. (eds) (2001), *The Breakdown of Class Politics: A Debate on Post-Industrial Stratification*. Baltimore: John Hopkins University Press.

Colgan, F. and Ledwith, S. (eds) (2002), *Gender, Diversity and Trade Unions: International Perspectives*. London: Routledge.

Creswell, J. W. (1998), *Qualitative Inquiry and Research Design: Choosing Among Five Traditions*. London: Sage.

Crompton, R. (1996), 'The Fragmentation of Class Analysis', *The British Journal of Sociology*, 47 (1): 56–67.

Crompton, R. (1998), *Class and Stratification: An Introduction to Current Debates*. Cambridge: Polity Press.

Crompton, R. and Scott, J. (2000), 'Introduction: The State of Class Analysis', in R. Crompton, F. Devine, M. Savage and J. Scott (eds), *Renewing Class Analysis*. Oxford: Blackwell, 1–15.

Crompton, R. and Scott, J. (2005), 'Class Analysis: Beyond the Cultural Turn', in F. Devine, M. Savage, J. Scott and R. Crompton (eds), *Rethinking Class: Culture, Identities and Lifestyle*. Basingstoke: Palgrave Macmillan, 186–203.

Crompton, R., Devine, F., Savage, M. and Scott, J. (eds) (2000), *Renewing Class Analysis*. Oxford: Blackwell.

Crook, S., Pakulski, J. and Waters, M. (1992), *Postmodernization: Change in Advanced Society*. London: Sage.

Crossley, N. (1996), *Intersubjectivity: The Fabric of Social Belonging*. London: Sage.

Crossley, N. (2001), *The Social Body: Habit, Identity and Desire*. London: Sage.

Dennis, N., Henriques, F. and Slaughter, C. (1969), *Coal is Our Life: An Analysis of a Yorkshire Mining Community* (2nd edn). London: Tavistock.

Devine, F. (1998), 'Class Analysis and the Stability of Class Relations', *Sociology*, 32 (1): 23–42.

Devine, F. (2004), *Class Practices: How Parents Help Their Children Get Good Jobs*. Cambridge: Cambridge University Press.

Devine, F. and Savage, M. (2000), 'Conclusion: Renewing Class Analysis', in R. Crompton, F. Devine, M. Savage and J. Scott (eds), *Renewing Class Analysis*. Oxford: Blackwell, 184–99.

Devine, F. and Savage, M. (2005), 'The Cultural Turn, Sociology and Class Analysis', in F. Devine, M. Savage, J. Scott and R. Crompton (eds), *Rethinking Class: Culture, Identities and Lifestyle*. Basingstoke: Palgrave Macmillan, 1–23.

Devine, F., Savage, M., Scott, J. and Crompton, R. (eds) (2005), *Rethinking Class: Culture, Identities and Lifestyle*. Basingstoke: Palgrave Macmillan.

Doogan, K. (2009), *The New Capitalism? The Transformation of Work*. Cambridge: Polity.

Elder-Vass, D. (2007), 'Reconciling Archer and Bourdieu in an Emergentist Theory of Action', *Sociological Theory*, 25 (4): 325–46.

Elliott, A. (2002), 'Beck's Sociology of Risk: A Critical Assessment', *Sociology*, 36 (2): 293–315.

Emmison, M. and Western, M. (1990), 'Social Class and Social Identity: A Comment on Marshall et al.', *Sociology*, 24 (2): 241–53.

Esping-Anderson, G. (ed.) (1993), *Changing Classes: Stratification and Mobility in Post-Industrial Societies*. London: Sage.

Evans, G. (2007), *Educational Failure and Working Class White Children in Britain*. Basingstoke: Palgrave Macmillan.

Fairclough, N. (2000), *New Labour, New Language*. London: Routledge.

Fantasia, R. (1995), 'From Class Consciousness to Culture, Action, and Organization', *Annual Review of Sociology*, 21: 269–87.

Featherstone, M. (1991), *Consumer Culture and Postmodernism*. London: Sage.

Fenton, S. and Dermott, E. (2006), 'Fragmented Careers? Winners and Losers in Young Adult Labour Markets', *Work, Employment and Society*, 20 (2): 205–21.

Fevre, R. (2007), 'Employment Insecurity and Social Theory: The Power of Nightmares', *Work, Employment and Society*, 21 (3): 517–35.

Furlong, A. and Cartmel, F. (2007), *Young People and Social Change: New Perspectives* (2nd edn). Maidenhead: Open University Press.

Gallie, D. (1978), *In Search of the New Working Class*. Cambridge: Cambridge University Press.

Gallie, D. (2000), 'The Labour Force', in A. H. Halsey and J. Webb (eds), *Twentieth-Century British Social Trends*. Basingstoke: Macmillan, 281–323.

Gane, N. (2004), *The Future of Social Theory*. London: Continuum.

Gewirtz, S., Ball, S. J. and Bowe, R. (1995), *Markets, Choice and Equity in Education*. Buckingham: Open University Press.

Giddens, A. (1979), *Central Problems in Social Theory: Action, Structure and Contradiction in Social Analysis*. London: Macmillan.

Giddens, A. (1984), *The Constitution of Society: Outline of the Theory of Structuration*. Cambridge: Polity Press.

Giddens, A. (1990), *The Consequences of Modernity*. Cambridge: Polity Press.

Giddens, A. (1991), *Modernity and Self-Identity: Self and Society in the Late Modern Age*. Cambridge: Polity Press.

Giddens, A. (1994a), 'Living in a Post-Traditional Society', in U. Beck, A. Giddens and S. Lash, *Reflexive Modernization: Politics, Tradition and Aesthetics in the Modern Order*. Cambridge: Polity Press, 56–109.

Giddens, A. (1994b), 'Risk, Trust, Reflexivity', in U. Beck, A. Giddens and S. Lash, *Reflexive Modernization: Politics, Tradition and Aesthetics in the Modern Order*. Cambridge: Polity Press, 184–97.

Giddens, A. (1994c), *Beyond Left and Right: The Future of Radical Politics*. Cambridge: Polity Press.

Giddens, A. (1995), *A Contemporary Critique of Historical Materialism. Volume One: Power, Property and the State* (2nd edn). Basingstoke: Palgrave Macmillan.

Giddens, A. (1997), 'Risk Society: The Context of British Politics', in J. Franklin (ed.), *The Politics of Risk Society*. Cambridge: Polity Press, 23–34.

Giddens, A. (1998), *The Third Way: The Renewal of Social Democracy*. Cambridge: Polity Press.

Giddens, A. (2000), *The Third Way and Its Critics*. Cambridge: Polity Press.

Giddens, A. (2001), 'Introduction', in A. Giddens (ed.), *The Global Third Way Debate*. Cambridge: Polity Press, 1–24.

Giddens, A. (2002), *Runaway World: How Globalisation is Reshaping our Lives* (2nd edn). London: Profile.

Giddens, A. (2006), *Sociology* (5th edn). Cambridge: Polity Press.

Giddens, A. (2007), *Europe in the Global Age*. Cambridge: Polity Press.

Giddens, A. and Hutton, W. (2001), 'In Conversation', in W. Hutton and A. Giddens (eds), *On the Edge: Living with Global Capitalism*. London: Vintage, 1–51.

Goldthorpe, J. H. (with Llewellyn, C. and Payne, C.) (1980), *Social Mobility and Class Structure in Modern Britain*. Oxford: Clarendon Press.

Goldthorpe, J. H. (with Llewellyn, C. and Payne, C.) (1987), *Social Mobility and Class Structure in Modern Britain* (2nd edn). Oxford: Clarendon Press.

Goldthorpe, J. H. (1991), 'A Response', in J. Clark, C. Modgil and S. Modgil (eds), *John H. Goldthorpe: Consensus and Controversy*. London: The Falmer Press, 399–438.

Goldthorpe, J. H. (2000), *On Sociology: Numbers, Narratives, and the Integration of Research and Theory*. Oxford: Oxford University Press.

Goldthorpe, J. H. (2002), 'Globalisation and Social Class', *West European Politics*, 25 (3): 1–28.

Goldthorpe, J. H. (2007a), *On Sociology, Second Edition: Volume One: Critique and Program*. Stanford: Stanford University Press.

Goldthorpe, J. H. (2007b), *On Sociology, Second Edition: Volume Two: Illustration and Retrospect*. Stanford: Stanford University Press.

Goldthorpe, J. H. (2007c), 'Cultural Capital: Some Critical Observations', *Sociologica*, 2: doi: 10.2383/24755.

Goldthorpe, J. H. and Lockwood, D. (1964), 'Affluence and the British Class Structure', *The Sociological Review*, 11 (2): 133–63.

Goldthorpe, J. H. and McKnight, A. (2006), 'The Economic Basis of Social Class', in S. Morgan, D. B. Grusky and G. S. Fields (eds), *Mobility and Inequality: Frontiers of Research from Sociology and Economics*. Stanford: Stanford University Press, 109–36.

Goldthorpe, J. H. and Marshall, G. (1992), 'The Promising Future of Class Analysis: A Response to Recent Critiques', *Sociology*, 26 (3): 381–400.

Goldthorpe, J. H., Lockwood, D., Bechhofer, F. and Platt, J. (1968a), *The Affluent Worker: Industrial Attitudes and Behaviour*. Cambridge: Cambridge University Press.

Goldthorpe, J. H., Lockwood, D., Bechhofer, F. and Platt, J. (1968b), *The Affluent Worker: Political Attitudes and Behaviour*. Cambridge: Cambridge University Press.

Goldthorpe, J. H., Lockwood, D., Bechhofer, F. and Platt, J. (1969), *The Affluent Worker in the Class Structure*. Cambridge: Cambridge University Press.

Gorz, A. (1982), *Farewell to the Working Class*. London: Pluto Press.

Granovetter, M. (1973), 'The Strength of Weak Ties', *The American Journal of Sociology*, 78 (6): 1360–80.

Gray, J. (1998), *False Dawn: The Delusions of Global Capitalism*. London: Granta Books.

Gurwitsch, A. (1964), *The Field of Consciousness*. Pittsburgh: Duquesne University Press.

Habermas, J. (1987), *The Theory of Communicative Action, Volume Two: Lifeworld and System: A Critique of Functionalist Reason*. Cambridge: Polity Press.

Hall, S. and Jacques, M. (eds) (1989), *New Times: The Changing Face of Politics in the 1990s*. London: Lawrence and Wishart.

Hall, S. and Jefferson, T. (eds) (1978), *Resistance Through Rituals*. London: Hutchinson.

Harvey, D. (1989), *The Condition of Postmodernity*. Oxford: Blackwell.

Hatcher, R. (1998), 'Class Differentials in Education: Rational Choices?', *British Journal of Sociology of Education*, 19 (1): 5–24.

Heath, A., Martin, J. and Elgenius, G. (2007), 'Who Do We Think We Are? The Decline of Traditional Identities', in A. Park, J. Curtice, K. Thomson, M. Phillips and M. Johnson (eds), *British Social Attitudes: The 23rd Report – Perspectives on a Changing Society*. London: Sage, 1–34.

Hebdige, D. (1979), *Subculture: The Meaning of Style*. London: Methuen.

Hebson, G. (2009), 'Renewing Class Analysis in Studies of the Workplace', *Sociology*, 43 (1): 27–44.

HEFCE (2005), *Young Participation in Higher Education*. London: HEFCE.

Hindess, B. (1987), *Politics and Class Analysis*. Oxford: Blackwell.

Hobsbawm, E. (1981), 'The Forward March of Labour Halted?', in M. Jacques and F. Mulhern (eds), *The Forward March of Labour Halted?* London: Verso, 1–19.

Hoggart, R. (1957), *The Uses of Literacy*. Harmondsworth: Penguin.

Holt, D. B. (1997), 'Distinction in America? Recovering Bourdieu's Theory of Tastes from its Critics', *Poetics*, 25: 93–120.

Howard, C. (2007), 'Three Models of Individualized Biography', in C. Howard (ed.), *Contested Individualization: Debates About Contemporary Personhood*. New York: Palgrave Macmillan, 25–43.

Husserl, E. (1973), *Experience and Judgement: Investigations in a Genealogy of Logic*. Evanston, Illinois: Northwestern University Press.

Husserl, E. (1976), *Ideas Pertaining to a Pure Phenomenology and to a Phenomenological Philosophy: Book 1*. The Hague: Martinus Nijhoff.

Husserl, E. (1977), *Cartesian Meditations: An Introduction to Phenomenology*. The Hague: Martinus Nijhoff.

Hyman, H. H. (1954), 'The Value Systems of Different Classes', in R. Bendix and S. M. Lispet (eds), *Class, Status and Power*. New York: The Free Press, 426–42.

Inglehart, R. (1977), *The Silent Revolution: Changing Values and Political Styles Among Western Publics*. Princeton: Princeton University Press.

Inglehart, R. (1990), *Culture Shift in Advanced Industrial Society*. Princeton: Princeton University Press.

Jackson, B. (1972), *Working Class Community*. Harmondsworth: Penguin.

Jackson, B. and Marsden, B. (1962), *Education and the Working Class*. London: Routledge and Kegan Paul.

Jameson, F. (1991), *Postmodernism, or, The Cultural Logic of Late Capitalism*. London: Verso.

Jenkins, R. (1996), *Social Identity*. London: Routledge.

Jessop, B. (1974), *Traditionalism, Conservatism and British Political Culture*. London: Allen & Unwin.

Kapitzke, C. (2000), 'Information Technology as Cultural Capital', *Education and Information Technologies*, 5 (1): 49–62.

Kelly, K. and Richardson, M. (1996), 'The Shaping of the Bristol Labour Movement. 1885–1985', in M. Dresser and P. Ollerenshaw (eds), *The Making of Modern Bristol*. Tiverton: Redcliffe Press, 210–36.

Kingston, P. W. (2000), *The Classless Society*. Stanford: Stanford University Press.

Kumar, K. (1995), *From Post-Industrial to Post-Modern Society*. Oxford: Blackwell.

Laclau, E. and Mouffe, C. (1985), *Hegemony and Socialist Strategy: Towards a Radical Democratic Politics*. London: Verso.

Lahire, B. (2003), 'From the Habitus to an Individual Heritage of Dispositions: Towards a Sociology at the Level of the Individual', *Poetics*, 31 (5): 329–55.

Lahire, B. (2004), *La Culture des Individus*. Paris: La Découverte.

Lahire, B. (2005), *Portraits Sociologiques*. Paris: Armand Colin.

Lahire, B. (2008), 'The Individual and the Mixing of Genres: Cultural Dissonance and Self-Distinction', *Poetics*, 36: 166–88.

Lamont, M. (1992), *Money, Morals and Manners: The Culture of the French and the American Upper-Middle Class*. Chicago: University of Chicago Press.

Lamont, M. (2000), *The Dignity of Working Men*. Cambridge, Mass.: Harvard University Press.

Lareau, A. (2000), *Home Advantage: Social Class and Parental Intervention in Elementary Schooling* (2nd edn.). Lanham, Maryland: Rowman and Littlefield.

Lareau, A. (2003), *Unequal Childhoods: Class, Race and Family Life*. Berkeley: University of California Press.

Lash, S. (1990), *Sociology of Postmodernism*. London: Routledge.

Lash, S. (1994), 'Reflexivity and its Doubles: Structure, Aesthetics, Community', in U. Beck, A. Giddens and S. Lash, *Reflexive Modernization: Politics, Tradition and Aesthetics in the Modern Order*. Cambridge: Polity Press, 110–73.

Lash, S. (2002), 'Individualization in a Non-Linear Mode', in U. Beck, and E. Beck-Gernsheim, *Individualization: Institutionalized Individualism and its social and Political Consequences*. London: Sage, vii–xiii.

Lash, S. and Urry, J. (1987), *The End of Organized Capitalism*. Cambridge: Polity Press.

Lash, S. and Urry, J. (1994), *Economies of Signs and Space*. London: Sage.

Lau, R. W. K. (2004), 'Habitus and the Practical Logic of Practice: An Interpretation', *Sociology*, 38 (2): 369–87.

LeDoux, J. (2002), *Synaptic Self*. London: Penguin.

Lee, D. J. and Turner, B. S. (eds) (1996), *Conflicts About Class: Debating Inequality in Late Industrialism*. London: Longmann.

Lehmann, W. (2007), *Choosing to Labour? School-Work Transitions and Social Class*. Montreal: McGill-Queen's University Press.

Leisering, L. and Leibfried, S. (1999), *Time and Poverty in Western Welfare States: United Germany in Perspective*. Cambridge: Cambridge University Press.

Levitas, R. (2005), *The Inclusive Society? Social Exclusion and New Labour* (2nd edn). London: Palgrave Macmillan.

Lockwood, D. (1966), 'Sources of Variation in Working-Class Images of Society', *The Sociological Review*, 14 (3): 249–67.

Luckmann, T. (1983), *Life-Wold and Social Realities*. London: Heinemann.

McLennan, G. (1985), 'The Contours of British Politics: Representative Democracy and Social Class', in G. McLennan, D. Held and S. Hall (eds), *State and Society in Modern Britain*. Cambridge: Polity Press, 241–73.

McNay, L. (1999), 'Gender, Habitus and the Field: Pierre Bourdieu and the Limits of Reflexivity', *Theory, Culture & Society*, 16 (1): 95–117.

McNay, L. (2008), *Against Recognition*. Cambridge: Polity Press.

Mahony, P. and Zmroczek, C. (eds) (1997), *Class Matters: 'Working-Class' Women's Perspectives on Social Class*. London: Taylor and Francis.

Majima, S. and Savage, M. (2007), 'Have There Been Culture Shifts in Britain? A Critical Encounter with Ronald Inglehart', *Cultural Sociology*, 1 (3): 293–315.

Marshall, G. (1988), 'Some Remarks on the Study of Working-Class Consciousness', in D. Rose (ed.), *Social Stratification and Economic Change*. London: Hutchinson, 98–126.

Marshall, G. (1997), *Repositioning Class: Social Inequality in Industrial Societies*. London: Sage.

Marshall, G., Newby, H., Rose, D. and Vogler, C. (1988), *Social Class in Modern Britain*. London: Hutchinson.

Marx, K. (1968), 'The Eighteenth Brumaire of Louis Bonaparte', in *Selected Works in One Volume*. London: Lawrence and Wishart, 94–179.

Mercer, K. (1990), 'Welcome to the Jungle: Identity and Diversity in Postmodern Politics', in J. Rutherford (ed.), *Identity: Community, Culture, Difference*. London: Lawrence and Wishart, 43–71.

Merleau-Ponty, M. (2002), *Phenomenology of Perception*. London: Routledge.

Mikelson, R. A. (2003), 'Bourdieu, and the Anomaly of Women's Achievement Redux', *Sociology of Education*, 76 (4): 373–5.

Mitchell, J. (1983), 'Case and Situational Analysis', *The Sociological Review*, 31 (2): 187–211.

Moerman, M. (1988), *Talking Culture*. Philadelphia: University of Pennsylvania Press.

Morris, L. and Scott, J. (1996), 'The Attenuation of Class Analysis: Some Comments on G. Marshall, S. Roberts and C. Burgoyne, "Social Class and the Underclass in Britain and the United States"', *The British Journal of Sociology*, 47 (1): 45–55.

Moustakas, C. (1994), *Phenomenological Research Methods*. London: Sage.

Mythen, G. (2005a), 'Employment, Individualization and Insecurity: Rethinking the Risk Society Perspective', *The Sociological Review*, 53 (1): 129–49.

Mythen, G. (2005b), 'From Goods to Bads? Revisiting the Political Economy of Risk', *Sociological Research Online*, 10 (3): http://www.socresonline.org.uk/10/3/mythen.html

Natanson, M. (1974), *Phenomenology, Role and Reason*. Springfield, Illinois: Charles C. Thomas.

Noble, G. and Watkins, M. (2003), 'So, How Did Bourdieu Learn to Play Tennis?' *Cultural Studies*, 17 (3): 520–39.

Nollmann, G. and Strasser, H. (2007), 'Individualization as an Interpretive Scheme of Inequality: Why Class and Inequality Still Persist', in C. Howard (ed.), *Contested Individualization: Debates About Contemporary Personhood*. New York: Palgrave Macmillan, 81–97.

OECD (2008), *Education at a Glance*. Paris: OECD.

Offe, C. (1985), *Disorganized Capitalism: Contemporary Transformations of Work and Politics*. Cambridge: Polity Press.

ONS (2006), *Social Trends No. 36*. London: Office for National Statistics.

ONS (2009), *Social Trends No. 39*. London: Office for National Statistics.

Ostrow, J. M. (1990), *Social Sensitivity*. New York: SUNY Press.

Pahl, R. E. (1989), 'Is the Emperor Naked? Some Questions on the Adequacy of Sociological Theory in Urban and Regional Research', *International Journal of Urban and Regional Research*, 13 (4): 707–20.

Pahl, R. E. (1993), 'Does Class Analysis Without Class Theory Have a Promising Future? A Reply to Goldthorpe and Marshall', *Sociology*, 27 (2): 253–8.

Pakulski, J. and Waters, M. (1996), *The Death of Class*. London: Sage.

Panel on Fair Access to the Professions (2009), *Unleashing Aspiration*. London: Cabinet Office.

Parkin, F. (1968), *Middle-Class Radicalism*. Manchester: Manchester University Press.

Payne, G. and Williams, M. (2005), 'Generalization in Qualitative Research' *Sociology*, 39 (2): 295–314.

Peterson, R. A. (1992), 'Understanding Audience Segmentation: From Elite and Mass to Omnivore and Univore', *Poetics*, 21: 243–58.

Phillips, T. and Western, M. (2005), 'Social Change and Social Identity: Postmodernity, Reflexive Modernisation and the Transformation of Social Identities in Australia', in F. Devine, M. Savage, J. Scott and R. Crompton (eds), *Rethinking Class: Culture, Identities and Lifestyle*. Basingstoke: Palgrave Macmillan, pp. 163–85.

Popper, K. R. (2002), *The Logic of Scientific Discovery*. London: Routledge.

Porter, S. (2002), 'Critical Realist Ethnography', in T. May (ed.), *Qualitative Research in Practice*. London: Sage, 53–72.

Potter, G. (2000), 'For Bourdieu, Against Alexander: Reality and Reduction', *Journal for the Theory of Social Behaviour*, 30 (2): 229–46.

Prieur, A., Rosenlund, L. and Skjott-Larsen, J. (2008), 'Cultural Capital Today: A Case Study from Denmark', *Poetics*, 36: 45–71.

Prior, N. (2005), 'A Question of Perception: Bourdieu, Art and the Postmodern', *British Journal of Sociology*, 56 (1): 123–39.

Pultzer, P. (1972), *Political Representation and Elections in Britain*. London: Allen and Unwin.

Reay, D. (1998a), *Class Work: Mothers' Involvement in their Children's Primary Schooling*. London: UCL Press.

Reay, D. (1998b), 'Rethinking Social Class: Qualitative Perspectives on Class and Gender', *Sociology*, 32 (2): 259–75.

Reay, D. (2005), 'Doing the Dirty Work of Social Class? Mother's Work in Support of their Children's Schooling', *The Sociological Review*, 53 (s. 2): 104–15.

Reay, D., David, M. E. and Ball, S. (2005), *Degrees of Choice: Social Class, Race and Gender in Higher Education*. Stoke-on-Trent: Trentham Books.

Reid, I. (1998), *Class in Britain*. Cambridge: Polity Press.

Rimmer, M. (forthcoming), 'Beyond Omnivores and Univores', *Cultural Sociology*.

Roberts, B. (2002), *Biographical Research*. Buckingham: Open University Press.

Roberts, K. (2001), *Class in Britain*. Basingstoke: Palgrave Macmillan.

Roberts, K. (2004), 'Leisure Inequalities, Class Divisions and Social Exclusion in Present-Day Britain', *Cultural Trends*, 13 (2): 57–71.

Roberts, K., Clark, S. C. and Wallace, C. (1994), 'Flexibility and Individualisation: A Comparison of Transitions into Employment in England and Germany', *Sociology*, 28 (1): 31–54.

Roemer, J. E. (1982), *A General Theory of Exploitation and Class*. Cambridge, Mass.: Harvard University Press.

Roker, D. (1993), 'Gaining the Edge: Girls at a Private School', in I. Bates and G. Riseborough (eds), *Youth and Inequality*. Buckingham: Open University Press, 122–38.

Särlvik, B. and Crewe, I. (1983), *Decade of Dealignment*. Cambridge: Cambridge University Press.

Saunders, P. (1987), *Social Theory and the Urban Question* (2nd edn). London: Routledge.

Savage, M. (2000), *Class Analysis and Social Transformation*. Buckingham: Open University Press.

Savage, M. (2003), 'A New Class Paradigm?', *British Journal of Sociology of Education*, 24 (4): 535–41.

Savage, M. (2005), 'Working-Class Identities in the 1960s: Revisiting the Affluent Worker Study', *Sociology*, 39 (5): 929–46.

Savage, M. (2007), 'Changing Social Class Identities in Post-War Britain: Perspectives from Mass Observation', *Sociological Research Online*, 12 (3): http://www.socresonline.org.uk/12/3/6.html

Savage, M. (2009), 'Contemporary Sociology and the Challenge of Descriptive Assemblage', *European Journal of Social Theory*, 12 (1): 155–74.

Savage, M. and Burrows, R. (2007), 'The Coming Crisis of Empirical Sociology', *Sociology*, 41 (5): 885–99.

Savage, M., Bagnall, G. and Longhurst, B. (2001), 'Ordinary, Ambivalent and Defensive: Class Identities in the Northwest of England', *Sociology*, 35 (4): 875–92.

Savage, M., Bagnall, G. and Longhurst, B. (2005a), *Globalization and Belonging*. London: Sage.

Savage, M., Bagnall, G. and Longhurst, B. (2005b), 'Local Habitus and Working Class Culture', in F. Devine, M. Savage, J. Scott and R. Crompton (eds), *Rethinking Class: Culture, Identities and Lifestyle*. Basingstoke: Palgrave Macmillan, 95–122.

Savage, M., Warde, A. and Devine, F. (2005c), 'Capitals, Assets and Resources: Some Critical Issues', *The British Journal of Sociology*, 56 (1): 31–47.

Savage, M., Barlow, J., Dickens, P. and Fielding, A. J. (1992), *Property, Bureaucracy and Culture: Middle-Class Formation in Contemporary Britain*. London: Routledge.

Sayer, A. (2002), 'What Are You Worth? Why Class is an Embarrassing Subject', *Sociological Research Online*, 7 (3): http://www.socresonline.org.uk/7/3/sayer.html

Sayer, A. (2005), *The Moral Significance of Class*. Cambridge: Cambridge University Press.

Schutz, A. (1962), *Collected Papers, Volume 1: The Problem of Social Reality*. The Hague: Martinus Nijhoff.

Schutz, A. (1970a), *On Phenomenology and Social Relations*. Chicago: University of Chicago Press.

Schutz, A. (1970b), *Reflections on the Problem of Relevance*. New Haven: Yale University Press.

Schutz, A. (1972), *The Phenomenology of the Social World*. Evanston, Illinois: Northwestern University Press.

Schutz, A. and Luckmann, T. (1973), *The Structures of the Life-World* (vol. 1). London: Heinemann.

Schutz, A. and Luckmann, T. (1989), *The Structures of the Life-World* (vol. 2). Evanston, Illinois: Northwestern University Press.

Schroeder, A. (2009), 'Patterns of Social Mobility in an Individualized World', paper presented at the British Sociological Association annual conference, Cardiff, 17 April 2007.

Scott, J. (1991) *Who Rules Britain?* Cambridge: Polity.

Scott, J. (2002), 'Social Class and Stratification in Late Modernity', *Acta Sociologica*, 45 (1): 23–35.

Searle, J. R. (1983), *Intentionality*. Cambridge: Cambridge University Press.

Searle, J. R. (1992), *The Rediscovery of the Mind*. Cambridge, Mass.: MIT Press.

Silva, E. B. (2006), 'Homologies of Social Space and Elective Affinities: Researching Cultural Capital', *Sociology*, 40 (6): 1171–89.

Silva, E. B. and Warde, A. (eds) (2010), *Cultural Analysis and Bourdieu's Legacy: Settling Accounts and Developing Alternatives*. London: Routledge.

Sintas, J. L. and Álvarez, E. G. (2002), 'Omnivores Show Up Again: The Segmentation of Cultural Consumers in Spanish Social Space', *European Sociological Review*, 18 (3): 353–68.

Skeggs, B. (1997), *Formations of Class and Gender*. London: Sage.

Skeggs, B. (2004), *Class, Self, Culture*. London: Routledge.

Skeggs, B. and Wood, H. (2008), 'Spectacular Morality: Reality Television, Individualisation and the Remaking of the Working Class', in D. Hesmondhalgh and J. Toynbee (eds), *The Media and Social Theory*. London: Routledge, 177–93.

Skeggs, B., Thumin, N. and Wood, H. (2008), '"Oh Goodness, I *am* Watching Reality TV": How Methods Make Class in Audience Research', *European Journal of Cultural Studies*, 11 (1): 5–24.

Sørenson, A. B. (2000), 'Toward a Sounder Basis for Class Analysis', *The American Journal of Sociology*, 105 (6): 1523–58.

Southerton, D. (2002), 'Boundaries of "Us" and "Them": Class, Mobility and Identification in a New Town', *Sociology*, 36 (1): 171–93.

Steinbock, A. J. (1995), *Home and Beyond: Generative Phenomenology After Husserl*. Evanston, Illinois: Northwestern University Press.

Strangleman, T. (2007), 'The Nostalgia for Permanence at Work?', *Sociological Review*, 55 (1): 81–103.

Sullivan, A. (2007), 'Cultural Capital, Cultural Knowledge and Ability', *Sociological Research Online*, 12 (6): http://www.socresonline.org.uk/12/6/1.html

Surridge, P. (2007), 'Class Belonging: A Quantitative Exploration of Identity and Consciousness', *The British Journal of Sociology*, 58 (2): 207–26.

Sweetman, P. (2003), 'Twenty-First Century Dis-ease? Habitual Reflexivity or the Reflexive Habitus', *The Sociological Review*, 51 (4): 528–49.

Thompson, J. (1995), *The Media and Modernity*. Cambridge: Polity Press.

Titchen, A. and Hobson, D. (2005), 'Phenomenology', in B. Somekh and C. Lewin (eds), *Research Methods in the Social Sciences*. London: Sage, 121–30.

Threadgold, S. and Nilan, P. (2009), 'Reflexivity of Contemporary Youth, Risk and Cultural Capital', *Current Sociology*, 57 (1): 47–68.

Trow, M. (2005), 'Reflections on the Transition from Elite to Mass to Universal Access: Forms and Phases of Higher Education in Modern Societies since WWII', in P. Altbach (ed.), *International Handbook of Higher Education*. Kluwer, 343–80.

Vandenberghe, F. (1999), '"The Real is Relational": An Epistemological Analysis of Pierre Bourdieu's Generative Structuralism', *Sociological Theory*, 17 (1): 32–67.

Van Gyes, G., de Witte, H. and Pasture, P. (eds) (2001), *Can Class Still Unite?* Aldershot: Ashgate.

Vester, M. (2005), 'Class and Culture in Germany', in F. Devine, M. Savage, J. Scott and R. Crompton (eds), *Rethinking Class: Culture, Identities and Lifestyle*. Basingstoke: Palgrave Macmillan, 69–94.

Vincent, C. and Ball, S. J. (2007), '"Making Up" the Middle-Class Child: Families, Activities and Class Dispositions', *Sociology*, 41 (6): 1061–77.

Wacquant, L. J. D. (1989), 'Social Ontology, Epistemology and Class: On Wright's and Burawoy's Politics of Knowledge', *Berkeley Journal of Sociology*, 34: 165–86.

Wacquant, L. J. D. (1991), 'Making Class: The Middle Class(es) in Social Theory and Social Structure', in S. G. McNall, R. F. Levine and R. Fantasia (eds), *Bringing*

Class Back In: Contemporary and Historical Perspectives. Boulder, Colorado: Westview Press, 39–64.

Wacquant, L. J. D. (2004a), 'Decivilizing and Demonizing: The Social and Symbolic Remaking of the Black American Ghetto and Elias in the Dark Ghetto', in S. Loyal and S. Quilley (eds), *The Sociology of Norbert Elias*. Cambridge: Cambridge University Press, 95–121.

Wacquant, L. J. D. (2004b), *Body and Soul: Notebooks of an Apprentice Boxer*. New York: Oxford University Press.

Wacquant, L. J. D. (2005), 'Habitus', in J. Berckert and M. Zafirovski (eds), *Encyclopedia of Economic Sociology*. London: Routledge, 315–19.

Wacquant, L. J. D. (2008a), 'Taking Bourdieu into the Ghetto: Social Theory Meets Urban Outcasts', plenary lecture delivered at *Putting Bourdieu to Work IV: A Working Conference*, University of Manchester, 29 May 2008.

Wacquant, L. J. D. (2008b), *Urban Outcasts*. Cambridge: Polity Press.

Walkerdine, V., Lucey, H. and Melody, J. (1999), *Growing Up Girl: Psychosocial Explorations of Gender and Class*. Basingstoke: Palgrave Macmillan.

Warde, A. (1994), 'Consumption, Identity-Formation and Uncertainty', *Sociology*, 28 (4): 877–98.

Weininger, E. B. (2005), 'Foundations of Pierre Bourdieu's Class Analysis', in E. O. Wright (ed.), *Approaches to Class Analysis*. Cambridge: Cambridge University Press, 82–118.

Westergaard, J. (1992), 'About and Beyond the "Underclass": Some Notes on Influences of Social Climate on British Sociology Today', *Sociology*, 26 (4): 575–87.

Williams, M. (2000), 'Interpretivism and Generalization', *Sociology*, 34 (2): 209–24.

Williams, M. (2002), 'Generalization in Interpretive Research', in T. May (ed.), *Qualitative Research in Action*. London: Sage, 125–43.

Willis, P. (1977), *Learning to Labour*. Farnborough: Saxon House.

Wood, E. M. (1986), *The Retreat From Class: A New 'True' Socialism*. London: Verso.

Wright, E. O. (1978), *Class, Crisis and the State*. London: New Left Books.

Wright, E. O. (1979), *Class Structure and Income Determination*. New York: Academic Press.

Wright, E. O. (1985), *Classes*. London: Verso.

Wright, E. O. (ed.) (2005), *Approaches to Class Analysis*. Cambridge: Cambridge University Press.

Wright, E. O. et al. (1989), *The Debate on Classes*. London: Verso.

Wuggenig, U. (2007), 'Pitfalls in Testing Bourdieu's Homology Assumptions Using Mainstream Social Science Methodology', *Poetics*, 35: 306–16.

Zweig, F. (1961), *The Worker in an Affluent Society: Family Life and Industry*. London: Heinemann.

Index